Packaging Materials

Ron Goddard

Pira
Randalls Road, Leatherhead
Surrey KT22 7RU
Tel: 0372 376161
Fax: 0372 377526

The facts set out in this publication are obtained from sources which we believe to be reliable. However, we accept no legal liability of any kind for the publication contents, nor for the information contained therein or conclusions drawn by any party from it.

No part of this publication may be reproduced, stored in a retrieval system, or transmitted, in any form or by any means, electronic, mechanical, photocopying, recording or otherwise without the prior permission, in writing, of the copyright holder.

© Pira, September 1990

Reprinted 1991, 1993

ISBN 0 902799 34 7

Published by

Pira – The Research Association for the Paper and Board, Printing and Packaging Industries

Printed in England by

Antony Rowe Ltd, Chippenham, Wiltshire

For sales and further information contact:

Marie Rushton

Manager, Information Services

Contents

Chapter 1 *The use of packaging materials and their selection*

INTRODUCTION   1

WORLDWIDE PATTERNS OF MATERIALS CONSUMPTION   4

REQUIREMENTS FOR SELECTION OF PACKAGING MATERIALS   6
Pack styles
Pack decoration
Methods of manufacture
Safety
Economics

FACTORS INFLUENCING THE CHOICE OF MATERIALS   8
Resources
Packaging design
Environmental aspects
Social changes
World technological developments
Processing developments
Materials substitution

Chapter 2 *Survey of traditionally-used materials*

WOOD   17

GLASS   19

METALS   25
Tinplate
Aluminium

PAPER   35
Papermaking
Packaging papers
Synthetic papers

New developments
Cartonboard
Corrugated fibreboard
Solid fibreboard

Chapter 3 *Synthetic materials (plastics)*

INTRODUCTION	49
POLYOLEFINS	59
VINYL-BASED POLYMERS	67
STYRENIC PLASTICS	71
POLYESTERS	74
POLYAMIDES (NYLON)	82
HIGH BARRIER MATERIALS	86
SPECIAL HIGH PERFORMANCE MATERIALS	89
MISCELLANEOUS PLASTICS	93

Chapter 4 *Flexibles and composite materials*

REGENERATED CELLULOSE AND ITS DERIVATIVES   95
Cellulose derivatives

MODIFIED AND BLENDED PLASTICS   99
Multilayer structure and laminates

SURFACE COATINGS   102
Additives
Blends and alloys
Surface treatments

COMPOSITE STRUCTURES   114

Chapter 5 *Ancillary materials*

ADHESIVES	**119**
ADHESIVE TAPES	**120**
MISCELLANEOUS CHEMICALS	**121**
ATMOSPHERE-MODIFYING CHEMICALS	**123**
CUSHIONING	**125**
INKS	**126**
LABEL MATERIALS **Paper based** **Other materials**	**128**
DEGRADABLE MATERIALS	**132**
WATER SOLUBLE FILMS	**138**

Preface

The use of different materials is frequently taken as a way to characterise the development of society. The Stone, Bronze, and Iron ages have come and gone (the last one very nearly perhaps), each material technology making possible a further improvement in the quality of life.

Packaging, too reflects changes in materials development. Manufacturers adopt the best of the materials available, either alone or in combination, to provide this function - especially important in the distribution of foods, without which modern life would be barely sustainable.

Although all of the traditional materials are still used in many packaging situations, there is no doubt that materials science is entering a major new phase of development. Some would call this the 'Plastics Age', but that is an over-simplification. Based on new research technologies, scientists can now understand why individual materials behave in the way they do, and use this knowledge to develop completely new ranges. Thus, it might be more accurate to call this the 'Age of Synthetics' - but without the pejorative overtones sometimes associated with that word.

In this review we have not attempted to give detailed textbook data on all of the materials used in packaging. To do so would have made it far too long, and duplicated the work of better-qualified specialists. Instead it is hoped, by introducing most of the materials used, and discussing their position in today's technology and their relevance to packaging, to give a useful overview of the present state of development.

Brief references to many of the topical developments and research, as well as concepts which are not yet commercial are included to indicate the timing and direction of future trends. Perhaps the most appropriate quote to take as the raison d'etre for this is from Rudy Pariser, former Research Director of the Du Pont Company: "Today's advanced material is tomorrow's commodity".

Chapter 1 *The use of packaging materials, and their selection*

INTRODUCTION

Throughout the world large quantities of material are used for the production of packaging; one estimate produced in 1988 puts the figure at about 160 million tonnes. In addition to the materials actually appearing as the end product, further larger amounts of resources are employed in the extraction, purification and processing stages, and significant quantities of energy, mostly in the form of fossil fuel, are also used.

The bulk of packaging is used to contain, protect and identify its contents during a single journey, and hence finishes up in trade or domestic refuse. Environmental concern over the extent of this consumption is a major factor in the minds of materials technologists seeking to conserve materials and reduce the load on the disposal of municipal solid waste. It also makes good economic sense. There are a number of ways in which this reduction can be achieved.

Enhanced performance of the materials is one, allowing lower quantities to be used in the first place. This is being progressed in most other industries, especially in such mass-critical uses as aerospace. Packaging manufacturers can gain a spin-off benefit from work done in these more demanding areas. A second route is to combine the best performance aspects of a number of different materials, each contributing its special mix of properties to the final result. Examples such as fibreglass reinforced plastics and carbon fibre/ceramics composites, do not appear directly as packaging materials, although they are likely to help improve the economics in the manufacturing industries involved. More intimate mixtures such as metal alloys (now being joined by new polymer 'alloys') are making significant contributions. Where this multi-material science really comes into its own is in the package manufacturing area, and almost all permutations and combinations of the main material types (wood, paper, metals, glass and plastics) are employed. These are discussed in more detail in later chapters.

Of the four main groups of materials, paper and board and plastics have been showing most steady growth over recent years. In the graph below showing sales value (courtesy of Trade Indemnity plc), metal and glass are losing out to these two materials, especially when they are used in combination.

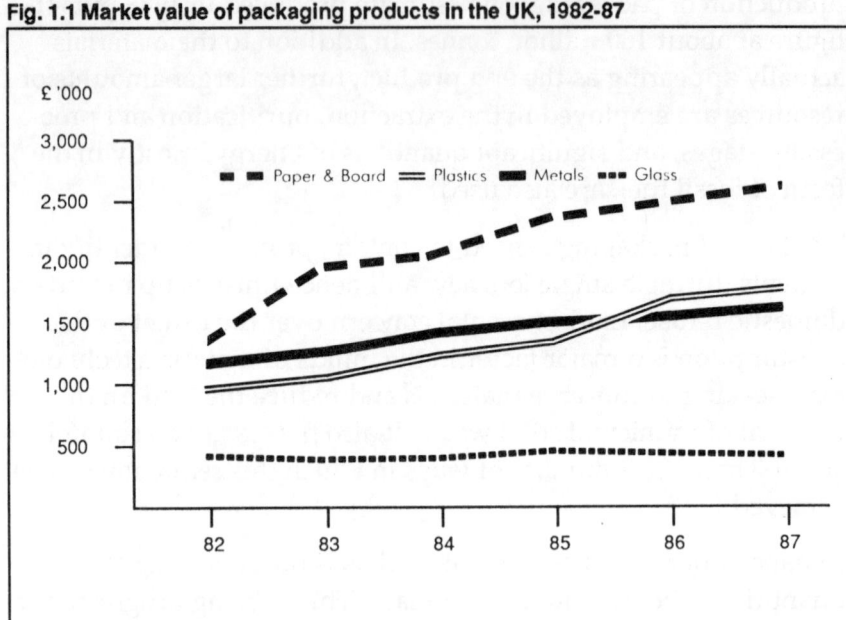

Fig. 1.1 Market value of packaging products in the UK, 1982-87

Important economies can also be achieved in the processing stages by improving materials utilisation, reducing losses due to faulty production, and making greater use of in-plant scrap. The importance of materials conservation varies between the different pack types. Metals are at the top of the scale – the materials account for between 75 and 80% of the final cost of a tinplate or aluminium container. Paper and plastics account for perhaps 50% of their total costs, and glass, with its cheap raw materials (essentially sand) is at the bottom with 20-25%.

All of these figures vary with the manufacturing process used and the level of technology involved. With metals and paper-based materials, in-plant scrap has to be collected and returned to the primary material producer. Plastics and glass manufacturers on the other hand, can usually put their scrap straight back into the process to produce new materials or containers. Already we have identified a conflict: the use of multi-material systems can seriously impair the economics of scrap re-use and, in many instances, of economic recovery by secondary processors.

Reducing the final volume of material for disposal is a different problem, and the most important primary decision – whether to make packs returnable/reusable or one-trip – does not concern us in this discussion. There is, however, a link with this decision since the materials manufacturer and the pack designer can greatly affect the ease with which pack or materials may be collected, identified and recycled to produce new packs or other useful items.

Another impetus for materials development is to keep up with demands for higher machine speeds at product manufacturers' plants. Sometimes an apparently minor constituent of a pack can make all the difference in this respect, the development of cold seal adhesive being one of the best examples. This allows horizontal form-fill machines to be run at very high speeds indeed, and these are now used extensively to pack confectionery countlines.

An initial look at the contents list for this volume is enough to show that materials development has been predominantly in the plastics sector. This is the youngest of the technologies and is still very much on its upward growth curve, taking us nearer to making true the statement that we are now in the plastics age. In mid-1988 the chief executive of Himont Plastics prophesied that by the year 2000 most of the steel used in consumable items (cars, household products, etc) will have been replaced by plastics, and that in a further 10 years most of the other metals will similarly have been displaced.

Certainly the properties of plastics have progressed a very long way since Mr Parkes stumbled across his synthetic ebony in the 19th Century, and most of this has occurred since the end of World War II. Plastics are now available which have the physical strength of steel, the temperature resistance of aluminium and the electrical conductivity of some metals. Although these specialised grades have been developed for engineering applications and hence are too expensive for packaging use, there is little doubt that some of the technology developed will allow their wider exploitation. New ideas and information may come from universities, research institutes and the R & D resources of the larger raw materials producers, and particularly as far as processing stages are concerned, of the large packaging manufacturing companies.

The spread of packaging consumption shown in Table 1.1 below is to some extent mirrored in the origin of innovation, although it must be said that the Japanese have taken packaging to their hearts to such an extent that they probably lead in the sheer 'novelty' stakes. Packaging is an international activity and the exchange of technical information has never been better, so it is not long before any significant development in one part of the world is known to interested practitioners in all others. In view of the vital role of packaging in improving the quality of life and reducing the losses of foods in developing countries, such international exploitation of improved technology is essential.

WORLDWIDE PATTERNS OF MATERIALS CONSUMPTION

Estimates of the world's consumption of packaging materials are necessarily very imprecise, and are likely to exclude many of the materials which are still used to a significant extent in certain areas of the world, eg re-used newspaper, and natural materials such as banana leaves. However, some attempts have been made to quantify the total. The three main consuming regions are the USA, Europe and Japan.

Table 1.1
Quantity and value of packaging materials used worldwide

Region	Total quantity in million tonnes	Total value £billion	Pop.(m)	per capita expenditure (£/annum)
USA	55	44	230	191
Europe	45	35	350	100
Japan	18	22	120	183
Rest of the world	42	14	4,300	3.2
Overall total	160	115	5,000	23

In the UK, the pattern of consumption of the different packaging materials indicates a steady, but not spectacular, growth in plastics over the past five years at the expense of all of the 'traditional' materials (metal, glass and paper).

Paper and board, still the largest single type of material, is holding its own to a greater extent than the other two, having lost only 1% of market share (2.4% of its size) over the period, while metal has fallen by 2.8% (one-eighth of its value) and glass by 1.5%, or a fifth, of its market value.

Approximate values for the major materials used in packaging, based on figures published by Rowena Mills Associates, which show these trends are as follows:

Table 1.2
UK consumption of packaging materials – trends 1984-88

Year	Paper Value £bn	Paper Share %	Plastics Value £bn	Plastics Share %	Metal Value £bn	Metal Share %	Glass Value £bn	Glass Share %	Total Value £bn
1984	2.14	41.2	1.49	28.7	1.18	22.1	0.38	7.3	5.19
1985	2.34	42.0	1.60	28.7	1.23	22.1	0.39	7.0	5.57
1986	2.45	41.9	1.78	30.4	1.24	21.2	0.38	6.5	5.85
1987	2.67	41.8	2.05	32.1	1.30	20.3	0.38	5.9	6.39
1988	2.83	40.4	2.38	34.0	1.39	19.9	0.40	5.7	7.00

REQUIREMENTS FOR SELECTION OF PACKAGING MATERIALS

Pack Styles

Materials are always going to be converted into a pack of some sort, either one that is supplied in a ready-made form or one which may be produced in-line on the filling machine, of which the most frequently seen are form-fill-seal types from flexible packaging, and cartons made from cartonboard reels or pre-cut blanks. There is, however, more scope for permutating materials types with different pack forms than might be at first thought. Some of the options are shown in Table 1.6 below. These apply to larger as well as retail packs.

Materials should always be selected on the basis of their performance. In considering the various materials, some will obviously be seen to be immediately unsuitable while others will have to be graded in terms of the degree of suitability. Features which are most important in the selection of packaging materials include:

> *A total non-interaction with the contents.*
>
> *Barrier properties to gases, light, water vapour and solvents.*
>
> *The ability to be heat-sealed.*
>
> *The capability to withstand high or low processing and filling temperatures.*
>
> *Mechanical strength, adequate for the particular application. Physical properties are likely to include rigidity, stiffness, tear resistance, puncture resistance and all of these over a varying range of climatic conditions which may be met with during storage, use or subsequent handling.*

Pack Decoration

Another element in the selection of materials is the form of decoration to be used. All too often this is left until the decision on the material type has been made, and then any inherent restrictions arising from that choice have to be dealt with.

Table 1.3

Decorating options for various packaging materials

Method of achieving decorative effect	Suitable for containers made from
Self-colour – integral in base material	Glass, plastics, paper
Embossed or moulded design	Plastics, metal, paper
Surface coated eg sprayed, metallised or dipped	Paper, plastics, glass
Surface printed	
a) prior to fabrication	Paper, plastics, metal
b) after fabrication into pack	Glass, metal, plastics, paper
Labels (paper or plastics) adhered or shrunk on	All forms of packaging materials
Labels adhered in forming mould	Plastics only

Methods of manufacture

Here again a wider range of manufacturing options may be possible than appears at first sight. The most commonly used forms for each of the materials are as follows:

> *Aluminium metal: Impact extrusion, deep drawing, cold-forming, spinning and, when rolled into thin foils, any flexible, or carton manufacturing process.*

> *Tinplate or tin-free steel: Rolling, bending and seaming (by many different techniques – some of which are discussed later), multi-stage stamping.*

> *Glass: Moulding direct from furnace into final shape, ribbon blowing.*

> *Paper and board: Moulding from pulp, corrugating, and conversion to bags and sacks (single or multi-ply), cartons or paper-based systems in widely different shapes, hot pressing, spiral winding, cut, crease and adhere or erect by mechanical interlocking. When paper components are combined with some plastics element, their versatility is vastly increased.*

> *Plastics: Extrusion or injection blow moulding, thermoforming, sheet fabrication and adhesion or heat/RF sealing, injection moulding, film production, laminating, fibrillation, weaving, perforating,etc, etc.*

Safety

All legal and safety requirements must be met by the material chosen for a particular use. These include safety aspects, eg for hazardous chemicals or other products, and hygiene and safety considerations for products which may leach their components into the packed food, or otherwise interact.

Economics

Finally and although mentioned last, clearly not least in importance, is the overall economics of the use of each material under consideration. It cannot be stressed too strongly that when considering the economics of a packaging material, pack or system, it is the *total* cost which must be considered. All too often, materials or packs are chosen on the basis of their unit price only. Elements such as associated packaging costs, storage, transit, protective requirements and the efficiency with which each is used on the intended line, can all have a major impact on the overall economics.

FACTORS INFLUENCING THE CHOICE OF MATERIALS

Resources

Historically man has used whatever materials which a) were available and b) he had the knowledge and technology to adapt. Thus packaging began with natural materials – gourds, animal skins, large leaves – progressed to easily-worked materials, mainly wood and clay, and then on to paper, metal and glass, and finally to plastics. The last group is quite different from any of the previously listed materials in that they are not a simple conversion of form of an existing material or mixture of materials, but involve the modification of the basic structures of chemicals to produce completely new compounds which do not exist in nature.

The pattern of development is, however, the same for each, from small scale discovery through experiment to advanced-scale manufacture, the cost falling at each stage.

Most of the materials used for early forms of packaging are either renewable (wood and vegetable fibre for boxes, and paper-based packaging) or very plentiful (sand for glass, clay for ceramics, iron ores for metal) and much later, bauxite for aluminium. There are exceptions to this claimed abundance, the tin used for tinplate is the most widely quoted. But suggestions for the period of time left before we will run out of this relatively scarce resource have been constant at "ten to twenty years" for the past twenty years, and there is currently no sign of it happening.

The other primary resource used for the manufacture of all forms of packaging is energy. Three main sources of this are used. One is the burning of fossil fuel – coal, peat, gas and oil. Another is the use of short term renewable forms of energy – mainly vegetable materials, wood and straw. The third is what might be called 'free' (although much debate can be expected on this definition). It includes solar, tidal, wind and hydro-electric energy. Geothermal (derived from the heat of the earth's core) and nuclear are not easily categorised. Nuclear can be called a form of fossil fuel consumption since the rare mineral (uranium) is a finite resource, but this is not the place to indulge in that debate.

Suffice it to say that all non-renewable resources based on minerals present in the earth can be recovered, however much they have been transformed during their use, provided that sufficient energy is available. This has led some commentators to suggest that in the long term the true cost of all forms of packaging (and of all other materials-consuming activities) will be determined by the amount of energy their manufacture and use employ.

Various attempts have been made to measure both the mineral and the energy consumption of packs, especially since environmental concern became important in the mid-1970s.

In 1989 Drs. Boustead and Hancock of the Open University, acting for the UK Government and industry, produced a four volume report on just one part of packaging consumption – that used for liquid foods. This, like all industrial activities, can be described in terms of an input of resources, their transformation into forms suitable for their intended use, and their discharge after use, back to the earth's reservoir (a concept called the "cradle to grave" approach). In this report all resources used were defined as being one of three types: renewable – mainly trees; non-renewable – fossil fuels being by far the most important; and minerals – natural compounds such as metal ores and water, plus oil used as feedstock. They studied this for 10 different types of container, and since the product packed also has a considerable influence on the packaging required, they examined a cross-section of 14 different liquid foods. Their analysis of the materials consumption expressed as the amounts used per million containers and per million litres (overall average) for the total of liquid foods, is given below.

Table 1.4
Gross energy and raw materials requirements of the total beverage system in the UK in 1986

Parameter	Units	Total for all containers delivered	Requirement per million containers delivered	Requirement per million litres delivered
Gross energy	MJ	1.47×10^{11}	5,216,137	8,963,330
Barytes	kg	4.14×10^5	14	25
Bauxite	kg	2.27×10^8	7,870	13,807
Sodium chloride	kg	7.28×10^8	25,246	44,291
Calcium sulphate	kg	8.73×10^6	303	531
Chalk	kg	3.41×10^5	12	21
Clay	kg	1.11×10^7	385	675
Feldspar	kg	2.86×10^7	992	1,739
Ferro-manganese	kg	6.02×10^5	21	37
Fluorspar	kg	4.54×10^6	157	276
Iron chromite	kg	7.69×10^5	27	47
Iron	kg	1.03×10^8	3,566	6,255
Lead	kg	5.53×10^4	2	3
Limestone	kg	4.74×10^8	16,420	28,807
Magnesium	kg	1.71×10^6	59	104
Manganese	kg	4.15×10^5	14	25
Metallurgical coal	kg	5.01×10^7	1,737	3,047
Natural gas feedstock	MJ	2.12×10^8	7,365	12,921
Oil feedstock	MJ	1.26×10^{10}	437,179	766,981
Rutile	kg	1.01×10^4	0	1
Sand	kg	7.69×10^8	26,643	46,743
Selenium	kg	1.45×10^4	1	1
Sodium nitrate	kg	2.90×10^4	1	2
Tin	kg	2.71×10^5	9	16
Water	gall*	1.93×10^{10}	670,523	1,176,356
Wood	kg	8.26×10^8	28,631	50,231
Zinc	kg	2.51×10^5	9	15

* 1 UK gallon = 4.55 litres

Energy requirements are better defined by the pack type, although since the container size makes an important difference it is often quoted in respect to a standard volume, usually one litre. Table 1.5 below provides the data in both forms.

Table 1.5

Energy consumption of different packs averaged over all delivery systems

Container system	Average volume ml	Average energy MJ[(1)]	Ranking	Average energy MJ[(2)]	Ranking
Brick cartons	587	3.700	1	6.302	1
Glass bottles	502	3.752	2	7.475	3
Gable top cartons	661	4.532	3	6.856	2
Tinplate cans	374	5.774	4	15.437	6
Aluminium cans	372	6.513	5	17.508	7
Other containers	1074	10.593	6	9.863	5
PET bottles	1511	13.740	7	9.093	4
Whole system	573	5.109		8.916	

(1) Calculated 'per container' (2) Calculated 'per litre of product'

Although the table above gives rankings, these should not be taken as giving anything more than a very generalised indication, since the difference between variants in each pack form will be very wide indeed (eg a 50ml PET spirits miniature compared to a 3l PET drinks bottle), and glass includes both returnable and single trip.

Packaging Design

This expression is frequently misunderstood, or treated in a very specific (graphics-related) way. Good design involves the selection of materials and their use in the most cost effective way to provide the best end result. First and foremost the item (pack in this instance) must be suitable for its purpose. It is surprising how many times this is not even included in a check list of design parameters. The selection of materials and form of packaging for any specific products are closely related. With a few exceptions it is in fact possible to use a number of materials to produce any pack form and, again with some exceptions, to use many materials to contain a product. One attempt to indicate the different packaging forms for which each material is suitable is given below.

Table 1.6
Materials available, and the forms into which they can be fabricated, using the coding below, are:

Material	Possible Forms											
	1	2	3	4	5	6	7	8	9	10	11	12
Aluminium	x	x	x	x	x	x			x			x
Glass	x			x	x							
Tinplate/Tin-free steel	x	x	x	x	x							
Cartonboard/corrugated		x	x	x		x		x			x	x
Paper		x	x	x		x		x		x	x	
Plastics	x	x	x	x	x	x	x	x	x	x	x	x
Flexible laminate		x	x	x		x	x	x	x	x		x

Forms
1   Bottles and jars (not precisely defined, the difference is mainly one of geometry and aperture size)
2   Open top cylinder cans and drums.
3   Square or parallelepipedal (brick shaped) packs
4   Irregular shaped three dimensional packs
5   Aerosol and other pressure dispense packs
6   Bags, sachets and sacks
7   Chub (large sausage-shaped containers)
8   Blister, skin and similar carded packs
9   Collapsible tubes
10 Heavy duty sacks
11 Transit cases and trays
12 Intermediate bulk containers

This first attempt at suggesting properties and their suitability for different packaging forms is based mainly on single material types. In practice, and increasingly commonly so today, materials are being combined to offer the best combination of their individual properties. Thus aluminium foil as a flexible component provides an absolute barrier to gas and water vapour. Polyethylene provides a heat sealing medium as well as increasing the mechanical strength and filling any tiny pinholes which may be present in the foil. Paper provides a low-cost form of mechanical stiffness and an excellent printing surface. Those three materials, paper, metal and plastics, put together in varying proportions provide some of the best known flexible materials, the laminate used for whole ranges of convenience foods. As is obvious from the grid of materials and suitability above, only plastics feature in every single form. They are indeed the most versatile of materials as will be apparent when they are discussed in more detail.

The design should make optimum use of materials, be completely safe, provide the best possible protection to the contents, be easy to use, attractive to help promote sales, provide full information to the purchaser, be easy to dispose of after use, and allow all of these to be met in the most economical way. Establishing agreed criteria for these different aspects is not easy and measuring them even more difficult. The only appropriate comment to make here is that costs should always be compared on the same basis, and should include every element – not for instance just the pack itself.

Environmental Aspects

In the industrialised world in recent years criticisms have been levelled at many aspects of packaging and its supplying industry by certain ecologists and environmental pressure groups. Their most frequent charges are that packaging a) makes excessive use of resources in a wasteful way, b) adds to the burden of waste disposal, and c) uses large amounts of energy including fossil fuel as feedstock for plastics. At different times other criticisms have been added including some from consumers, that the packaging adds unnecessarily to product costs, that it limits consumer choice, that it may be used in a deceptive manner, and that it often appears as a major constituent of litter. It has been recognised that there is substance in some of these claims, but they mainly arise from a misunderstanding of the role of packaging. Over-packaging is very difficult to define and it varies with time and place.

A number of attempts have been made to set out a code of good packaging, designed to meet the substance of these charges. One, originating from Japan in the early 1970s, was introduced in the UK. Its provisions (put very briefly) were that packaging must:

Meet all legal requirements.

Make economical use of materials.

Fully protect the contents.

Have no interaction between pack and content.

Be no larger than strictly necessary.

Be convenient in use.

Present all the required information.

Be designed with due regard to its possible effect on the environment.

The approach is, as can be seen, one of giving general guidance. Any more detailed strictures are neither practicable nor would they be acceptable. Different versions have been produced in other countries. A Swiss proposal called "The Ten Commandments" issued in 1989 calls for a similar set of conditions to be met.

In recent years, particularly since the mid-1980s, there has been a growing call for all products, including packaging, to be seen to be 'green' (ie sensitive to the ecological needs of the planet). Some individual companies and trade associations have attempted to exploit this concern, claiming better 'green' credentials for their product or material, but the issue is far from simple and until some objective form of environmental assessment is agreed, the whole area is very subjective. Various attempts to rate products, including packaging, by awarding some form of 'Eco-label' are in use or being studied in Canada, Japan, Norway, France, and particularly West Germany where the 'Blue Angel' labelling system has been in use for a number of years. One effect of what has been called "internecine warfare" between rival packaging materials suppliers, is to harden the attitude of the environmental protection groups against the whole industry.

Social Changes

The largest area of packaging is for foods, and the food manufacturers are most sensitive to changes in lifestyle and demography in their particular markets. There have been huge changes in social patterns in recent years and while these are at different stages in various parts of the world, the general trend is similar, albeit at different paces.

Social changes mean that more women have employment outside the home and hence demand greater convenience in food production and preparation. Eating habits have also changed significantly since the 1970s with less formal 'sit down' meals and more of what is called 'snacking'. Coupled with changes in resources inside the home (especially the dynamic growth of the microwave oven) this has led to a huge growth in demand for part- or fully-prepared meals and snacks. These in turn necessitate different forms of packaging, often involving very demanding performance.

World technological developments

Packaging is a truly international activity and although social, climatic and market situations differ widely, technology is readily transferable around the world. Those involved in pure packaging innovations and the application of these to different situations consider the world as a source of inspiration and information. Many would agree that Japan represents the major centre of sheer innovation, while the USA and Europe concentrate on fewer but larger scale developments.

In Japan, packaging is both an art form and a science but due to their cultural background consumers seem to especially appreciate two major features. One is very high quality and appearance – this comes from a long tradition of gift giving, with the presentation as important as the item itself. The other is in pure novelty, often for its own sake, and manufacturers will offer a bewildering range of packs to tempt the purchaser into buying consumer goods, especially where competitive market pressures are high as in the beer and soft drinks sectors.

Whilst other markets may not require this level of sophistication – or may consider it to be positively wasteful – the technical skills developed in meeting some of these design briefs can be more widely used in other packaging situations.

Processing Developments

The properties of materials can be much influenced by their processing, and hence this can be an important influence, amply demonstrated for instance by the difference between wrought and cast iron, and high tensile steel.

Today, with the wide range of basic materials available for packaging, which can be used either alone or in combination, modifying their nature by some processing technique may sometimes provide even broader spectra of properties. The treatment may be carried out at the time of manufacture or as a post-fabrication operation. Examples include orientation (stretching), surface treatment, foaming, coating, cross-linking by irradiation, crystallisation, cold rolling, and annealing as well as blending and combining by various means.

Materials substitution

As already stated, packaging is a constantly changing scene and its development has paralleled the changes in materials technologies. For that reason every new material is seen as a potential substitute for one or more of the existing types in use. Sometimes this is a perfect match and substitution rapidly occurs. Oriented polypropylene film which rapidly replaced much of the regenerated cellulose market is an example of this. In other situations there are limitations which make only partial substitution feasible – plastics to replace glass and metal for processed foods is a good example here.

When this competitive situation arises it provides a spur to development of both the new and the old materials. Users wishing to consider a substitute material have the option of demanding absolute comparability or adapting their application to make best use of the actual properties of the new material. Commonsense should be employed. For instance, glass is stable up to nearly 800°C, but such conditions are hardly likely to be met in actual use, so matching this would not be required of any alternative material.

Chapter 2 *Survey of traditionally-used materials*

WOOD

With the exception of clay and natural materials such as reeds and leaves, wood is the oldest material used for the manufacture of packaging. Examples of chests have been found in tombs of the old kingdom in Egypt, almost 5,000 years old. They were not in such good condition as those found in Tutankamun's tomb, but these had had much better storage conditions and are about 1,600 years younger.

Only in Japan is wood used to any significant extent for retail packs today, even the traditional cedar wood cigar boxes are being replaced by structural foamed plastics. Most applications in the western world are for heavy duty transit and industrial packaging, where the high stiffness, low weight and versatile construction options of wood can be best employed.

Despite the undoubted advantages which wood can offer, it has also some serious limitations and it is mainly in attempts to resolve these that research effort has been directed in the last decades. Moisture sensitivity is one such limitation, of particular importance if, for instance, a wooden pallet is sealed in with machinery inside a moisture barrier wrapping. The pallet may hold up to its own weight in water, enough to swamp the capacity of most desiccant packs which may be included. Impregnation with resins can reduce this moisture sensitivity, but the rate of penetration into large blocks of solid wood is very slow, and here composite structures are more frequently used.

The directional strength (and conversely in the opposite direction, weakness) of thin sheets have for years been compensated for by cross-laminating (plywood). This has extremely high stiffness and puncture resistance, it is of light weight, and can be joined by all types of fixing, from adhesives to plastic clips and nails. Plywood is available in many grades and thicknesses and is relatively inexpensive. Packs are frequently made from this using collapsible corner angles (riveted tinplate), and patented metal or plastics joining strips. Their ability to be collapsed and delivered flat gives these cases a great advantage over traditional rigid wooden versions.

Heavier crates for machinery, etc, often incorporate plywood panels framed in solid wood. A plus for many situations where timber crates are used is the re-use facility, an especial benefit in developing countries or where timber is a scarce resource. Hardboard is another wood-derived material which has some distinct advantages. Being made from a much larger proportion of the tree, and on a continuous process, it is more economic to produce. Since some binding component must be incorporated to hold the particles together, this can double as a moisture-resisting and even decorative finish. The material is available in various grades and in thicknesses from 2-12mm, with different surface finishes and textures. Panels can be used as direct substitutes for plywood, upgrading the thickness if it is necessary to match the puncture resistance of plywood. In the 1970s FIDOR, the Fibreboard Industry Development Organisation, funded a project at the UK Packaging Research Institute to study ways of increasing the use of hardboard for packaging. Exploiting the material's high rigidity by incorporating stiffening panels in many corrugated board or flexible containers was one recommendation. Another avenue explored was the production of cylindrical drums using heat and steam to roll the board into a tube. Some drums made in this way are currently on the UK market but the concept has not so far proven to match the economics of the more traditional convolute paper-wound drums.

Finally the attraction of one-piece lay-flat cases offered the potential for really high volume application. The principle was easy to prove with laboratory-made containers, and the performance of prototype styles was established to be very good, but the various ways of making the joints between panels (adhesive tapes, stitching, hinged channel extrusions, metal clips) were not yet at the right level of development to make this a viable way of producing on a large scale. The latest approach has been to develop a creasing technique analogous to that used in cartons and case making, initiating partial delamination of the built-up layers in the crease area. Some success has been reported but it is not yet considered likely to lead to very extensive manufacture of such cases.

More recently a process to give hardboard greater flexibility, ie make it 'tougher' rather than just strong, has been developed. The process involves the blending in of elastomeric materials such as rubber.

Another form in which wood is finding packaging applications is a reconstituted sheet material produced from wood chips and shavings. 'Sterling' board is one example, currently being used for pallet boards and case panels. Because this type of material is blended, conditioned, mixed with a binding agent and compressed under heat, it is possible to produce shaped items in a single process. Nestable pallets are a particularly appropriate example.

Other wood-like fibres can be used to produce rigid panels. These range from residues of sugar cane to straw and plant stems. The natural bonding ability of cellulose fibres, use of which is made in the production of paper, cannot be relied upon to form a strong material with these fibres so organic resins are usually incorporated.

One example, called Compak Board, is produced in a hot press operation from chopped straw – a material which is otherwise usually burned since it has no economic value. Other fibrous materials, eg rice or even rape stems as well as wood chips can also be incorporated. The properties can be varied according to the length and mix of fibres, proportion of resin and density of compaction. Some very high performance materials can be produced in this way. This is one of the materials marketed on its 'green' image, not only using a waste material but being degradable after use. BioPack, a form of cushioning made in Germany from chopped straw in paper bags, was introduced at the 1990 Interpack show.

GLASS

Glass has a pedigree almost as long as wood, but its widespread use for packaging goes back only about 200 years. The main ingredients (sand with smaller amounts of lime and other materials) have not changed over the centuries, but there has been a constant stream of technical improvements to the materials and to the processing.

The main constituent of glass is silica (sand) but modern glass contains a number of other minor ingredients to improve its melting ability, strength and appearance. A typical bottle grade might be produced from:

Table 2.1
Glass composition

Mineral	% by weight
Silicon oxide	70
Calcium oxide	10-12
Magnesium oxide	0.5-3
Sodium oxide	12-15
Alumina	1.5-2
Iron oxide	trace
Sulphur trioxide	trace

These are crushed, blended with about 20% of scrap glass (cullet) and fired in a furnace at about 1300°C. Higher levels of cullet can be used, and greater recycling through 'bottle banks' is making this possible. Coloured grades are made by the addition of traces of chrome oxide (green), cobalt oxide (blue) and iron plus sulphur (brown).

An invisible ingredient which is of particular importance for the glass industry is energy, and the amount of this needed has a big influence on the production economics. Among the techniques developed to reduce this energy consumption have been:

Pre-blending of ingredients and compaction of these into briquettes.

Arranging the raw material infeed to come "down the chimney" so that it becomes preheated before entering the furnace.

Computer control of ingredient mix, and the better distribution of this within the molten mass in the furnace.

Much improved temperature monitoring of the process and control of the annealing stage.

Greater use of cullet which, being already premixed and in a glassy state, requires considerably less energy to melt.

These developments relate to the production of raw material, but as already mentioned, the economics are equally affected by the associated stages of producing containers. First and foremost comes lightweighting – simply reducing the amount of glass in a container. The returnable one-pint milk bottle has been progressively reduced in weight from 600g in 1920 to 225g in 1989. As is well known, glass is an enormously strong material even in very thin sections (one has only to think of glass fibre reinforcing filaments, and the common electric light bulb) so it is theoretically possible to make very thin containers in glass. The weakness of glass is that it is brittle and tends to have its stresses locked into the surface layer. If this is damaged at all the material can be easily broken – which is how a glazier can snap 6mm glass between his fingers.

Much work has therefore gone into the four main routes to minimise this surface damage factor. One is to reduce internal stresses by improved annealing control. Since glass is a supercooled liquid it can cool at differential rates due to air currents, varying wall thickness, conductive effects from support surfaces and the proximity of other hot containers. Better understanding of these complex interacting effects (including the use of computer modelling techniques) has made possible major improvements.

A process which at first might be expected to lead to extra strains in the glass, using chilled air (at -196°C) to blow the bottles, has been developed by Swedish company AGA. It calls the process 'Cryo-glass Technology' and claims increased productivity and quality.

The second element is to ensure that the external surfaces have as few surface blemishes as possible by careful manufacture and handling of the moulds, and by equally careful handling of the glass parison and finished article during manufacture.

It is also important to consider this factor in the detailed design of packaging containers, especially if they are for multi-trip use. An analysis of the points of contact with guide rails, base supports, closure and labelling equipment on the intended filling lines makes it possible to build in 'wear absorbing' zones or strengthened sections. One design feature which allows for absorbing wear is to dimple the surface, usually around the thickest part of the girth but also on the underside of the base, where it may take the form of a series of raised radial lines. The high spots are rubbed away in a controlled manner with minimum effect on the overall strength. Some design limitations are inevitable in adopting this approach, but clever designers can achieve the effect while still providing good brand identity.

To supplement these inherent improvements there is a whole battery of surface coating options available which have two purposes – one, carried out on hot containers and hence referred to as 'hot-end treatment' to toughen the outer skin; the other, at the 'cold-end', to lubricate all external surfaces to reduce the potential damaging effects of glass to glass contacts which are impossible to avoid in the normal way of usage of glass containers. Hot-end treatments include the well known 'titanising' process in which a coating of titanium compounds is allowed to fuse into the surface.

A treatment originated in Japan by the Yamamura Glass Company is based on a more fundamental approach to the physical chemistry of glass. As the surface is a continuous layer of silicon, sodium and oxygen atoms, Yamamura reasoned that if the small diameter sodium ions can be replaced by larger potassium ions (which are chemically similar to sodium) this would result in a toughened surface skin to the glass. The beneficial effects have been confirmed in trials, and glass manufacturers in a number of countries have taken licences to use the process.

The capital cost of modifying the glass furnace is quite high, and so its practical applications have been slow to appear. The decision to invest the large sums involved depends upon the confidence of the industry in the future market and its growth. The glass container industry has been under considerable pressure from alternative materials – tinplate and aluminium in the beer market is one important competitor, and plastics for many ranges of dry and moist foods. The industry has put a great deal of effort and resources into encouraging the recycling of used glass bottles via bottle banks and the decline in consumption of glass containers has been slowed but not yet halted (see Table 1.2 on page 5).

Where surface layers of glass-compatible materials are fused on at high temperatures, the option also arises to make these of a different colour, either translucent or opaque. This treatment is carried out to great effect in Japan. It provides the flexibility for a glass manufacturer to offer a wide range of coloured glasses while maintaining the economics of scale and operational efficiency of an all-white glass production. In the UK, Rockware Glass has developed a new epoxy spray technique to achieve this effect. The incorporation of a 'fore-hearth' melting facility is one way that the problem of producing small quantities of "all-through" coloured glass, especially in pale shades, can be overcome. This is the route taken by United Glass, the UK arm of American company Owens-Illinois.

At the cold-end the most commonly used treatment is to coat with an emulsion of polyethylene and a silicone to provide the required degree of slip.

Another surface treatment, Vapocure, offers the possibility of a physical protection layer and a means of colouring plain glass bottles.

Some of the cold-end treatments are removed by the alkaline wash water used in returnable systems (for example, as used for beer and milk in the UK). A patent issued to Kirin Breweries describes a method of overcoming this by stripping away the old surface coating in alkaline wash water and recoating with a fresh layer after the washing stage. Thus each bottle leaves the filling line with a fresh and complete surface coating layer. It is claimed that the service life of bottles is significantly extended by this route. In 1990, GE Silicones introduced a similar concept to the UK specifically for returnable glass bottles.

Another Japanese-developed surface protective coating, from Yamamura Glass and Dainippon Inks, is a fluoracrylate/silane material which bonds into the glass surface.

Yet another from Japan, Mul-t-Cote from Star Chemical Company, involves dip coating with a double layer of plastics. The first is a 50μm coating of styrene butadiene rubber, followed by 20μm of a high modulus polyurethane. The thin top layer may be coloured if desired, but the main function is to prevent dangerous scattering of sharp fragments of glass if a pressurised bottle is dropped. Associated with glass in a special way is the development of protective sleeve labels made from heat shrinkable plastics. One system, Plastishield, makes use of a shrinkable sleeve of foamed polystyrene about 1mm thick. The printed quality of the later form of these is a big improvement on the earlier versions, and in addition to the main benefit of surface protection (including of the base rim), a number of other benefits can be claimed. These include reduced noise, some degree of thermal protection (to keep the contents hot or cold for longer) and a better and safer grip, especially if the surface is wet.

Offering some of the protective benefits of the Plastishield is the use of a shrink-sleeve of reverse-printed PVC film. These have been employed in Japan for about 15 years and more recently have achieved some rapid expansion into Europe and the UK. The label is usually printed, but can also include an overall colour to convert a clear glass bottle to an apparently coloured one, a technique which is extensively used in Japan. Recently, selective metallisation has also been added to the options. As well as protecting the container from surface damage by glass to glass contact, the sleeve can make a contribution to restricting the energy of flying fragments of glass in the event of a full or partly full bottle of a carbonated beverage being dropped and breaking. Perhaps one of the most unusual forms of these labels is the Japanese foil laminated example.

One of the most exciting areas of research is into novel ways of producing continuous layers of glass either in the form of three-dimensional vessels or as flat sheet. Work is being conducted on the first by International Partners in Glass Research in a long term project to see if the technique of 'cold precipitation', long known as a laboratory novelty, can be adapted to produce usable glass on a commercially viable scale. The principle is to take a solution of tetraethoxysilane and add water. Heating this at only 100-300°C has the effect of producing a glassy layer in situ. Initial applications have concentrated on very thin layers for microchips but it could have other potential in the long term future.

Similar in some respects (the build up of very thin layers in situ), but using a quite different technique, is the vapour deposition of silicon dioxide and other inorganic materials for the production of ultrathin layers of 'flexible glass' on plastics substrates. This technique is described in greater detail elsewhere.

A novel use proposed for scrap glass is the production of a lightweight 'foamed' glass material. The Nagoya Research Institute in Japan has been working on this, but no applications are yet noted.

One factor which may be very significant in the long term is the environmental concern which could bring about greater use of returnable bottles. Glass has some very significant advantages in this area and indeed is almost unique in its ability to be continuously recycled to a pure and inert form. This is technically straightforward, but depends very much upon the economics of recovery, sorting and transporting scrap glass. Another factor which could be of benefit for the glass industry is its suitability as a container for microwave cooking or reheating. In 1988 some shallow, lightweight glass bowls with heat sealed or crimped-on lids were offered for this market.

METALS

The most important metals used in packaging remain steel, tin and aluminium, the first overwhelmingly in the form of tinplate containers up to about 25l, and as steel plate for larger ranges of drums. Two other forms of coated steel plate are chrome/chrome oxide coated, and lacquered black plate, both often commonly referred to as tin-free steel. Small quantities of lead sheet are used for sealing tops of wine bottles but this is a dying practice on the grounds of health and costs.

Metals are also used in packaging for the manufacture of closures, strapping wire and tapes, open-top trays and as a thin foil in flexible packaging. In all of these the material is coming under increasing pressure from plastics materials. The balance of useful properties is broadly that the metals have the highest absolute performance in heat tolerance, physical strength, barrier, stiffness and deadfold, but plastics offer easier production routes to the finished items (such as closures) more complex designs, self-colouring, and heat sealability.

For different packaging applications the importance of each of these properties varies and hence the rate of substitution similarly differs. In a number of areas metal packaging is holding its own by being combined with other materials, especially plastics or paper-based components. The composite can with its modern derivatives is a good example.

Tinplate

Tinplate is made by hot rolling steel sheet to a number of standard thicknesses, most commonly 0.27-0.30µm, and then coating by electrolytic means with pure tin.

After trimming, most of the coiled reel stock is coated with an organic lacquer to further protect the steel against corrosion. Different lacquers have been developed for the packaging of various foods which vary in acidity. Tinplate need not be coated with the same weight of metal on each side and since the outside-facing surface is exposed to standard conditions irrespective of the contents, this can be of a standard grade. The pH value gives a measure of the acidity of a liquid, the lower numerical values being the most acid and the scale runs from 1 to 14 with 7 as the neutral point. The chromium/chromium oxide coated materials were developed in Japan during the 1960s at the time when tin prices looked set to rise dramatically. The material has two very thin layers, one of chromium metal (about $0.1g/m^2$) plus a coating of chromium oxide at about half this thickness. This compares with an average coating weight of tin of $5.6g/m^2$ (or over 50 times as much). Although the chromium/chromium oxide produces a bright metallic finish it does not give the same degree of protection against corrosion which tin provides, so it is essential that it is lacquered.

Lightweighting has also taken place steadily in the can industry. Since 1945 the weight of steel in a processed food can has been reduced by 35% and the tin by 80%. In modern cans the metallic tin layer accounts for only 0.4-0.5% by weight.

Developments in materials over recent years have included water-based lacquers, posing less environmental problems in their manufacture and use, and greater use of lacquered tin-free steel.

Iron foil has been produced commercially for many years and periodically its possible uses in packaging have been re-examined. The material can be produced by hot rolling, direct formation from the melt via a slit die, by a sintered powder compression route, or by electrolytic deposition. Its main drawback is that the very bright and reactive surface rusts easily and so needs immediate protection.

In the mid-1980s one development by Japanese company Toyo Seikan was successfully launched. Hiretoflex is a sandwich laminate of iron foil between two layers of very strongly bonded polyolefin materials. It can be formed into packs by folding, pressing and rolling into many different shapes, but the initial application was for the production of small portion packs to hold jellies. Toyo Seikan claims that the overall economics of manufacturing these packs from iron foil are better in terms of cost performance than if the much easier to work aluminium had been used.

At the UK Pakex Exhibition in 1989 Metalbox showed a similar concept called Metpolam. The metal may be either iron foil or aluminium and the initially projected application for this material was the production of aerosol container components where corrosion resistance is particularly important. These types of materials, based on coated metal foils, are likely to find increasing use in packaging since they offer the combination of total barrier, cold-formability, deadfold, are non-corrodable and can be heat sealed.

More obvious than the materials developments have been those connected with the manufacture of cans from tinplate. By far the most significant of these, and one which has found its greatest applications in the carbonated beverage and beer sector, is the two-piece can produced in place of the traditional three-piece item. While the latter is formed from sheet by rolling, side seaming and then seaming on the two ends, the two-piece type is stamped out of a round disc blank using very high pressures and sophisticated progressive dies. Two methods are used, known generically as draw-redraw (DRD), and draw-wall-iron (DWI).

The DWI technique involves stamping out a cup from sheet metal which has the same base as its intended final diameter. The material for the side is then obtained by stretching the short wall to the desired height. This becomes very thin, and work-hardened in the process, and the method is therefore used mainly to produce cans for carbonated beverages. DRD, a more versatile approach, uses more metal, since an oversized cup is first stamped out and this is then pressed by stages into deeper, narrower containers. As a result the walls are thicker and more uniform than DWI cans, but both methods result in considerable savings in material.

This saving of metal is of vital importance in manufacturing economics; as mentioned in the introduction, materials costs are a very high proportion of the total in canmaking, estimated at about 75% of the total cost. Naturally, with such a reduction in metal usage the wall thickness is reduced. The internal pressure of carbonated soft drinks (about 3.5 atmospheres, or $50lb/in^2$) keeps the wall of DWI cans completely rigid until opened.

The DRD can is being proposed for a much wider range of processed foods which are not themselves normally sealed under pressure. In order to introduce the necessary pressure, small volumes of liquid nitrogen are dropped into the can immediately before the lid is seamed on. This also displaces the headspace oxygen thus preventing oxidation spoilage. AGA in Sweden and Distillers MG of the UK offer complete systems for this.

Currently the two main non-beverage product sectors in which two-piece cans are used in the UK are processed pet foods and baby foods. Due to the large investment needed for high capacity production of two-piece containers, there is generally much less scope for size flexibility.

Forming techniques used for the side seam on three-piece cans, which are still by far the most commonly used for processed foods, have progressed from the early hooked and soldered type to induction welding. The benefits of the latter are such that it is now the main process for all high production plants. Material is saved since the overlap is reduced to below 1mm – a saving of metal which may be as high as 4% of the tinplate used. The weld is also stronger than the previously used soldered types, a fact which makes it particularly important for pressurised containers such as aerosols. The much smaller and very thin weld poses no problems for the top- and bottom-seaming operation unlike the earlier, bulky hooked side seams. Finally, there is is no need to leave a wide unprinted area for the wiped solder join, so all-round decoration is possible.

Two forms of electrical induction welding are used. The first, developed by the Continental Can Company in the USA, and called Conoweld, involves solid welding electrodes, but these can be sensitive to metal pick-up from the surfaces which impairs subsequent seals. It is therefore essential that very clean techniques are used. An alternative technique which gets round this problem is the Soudronic system, devised by Swiss company Soudronic AG. In this, a continuous length of copper wire is passed between the electrode and the material being welded. Since this is not re-used directly there is no problem with metal pick-up. Both processes have been developed to high speeds.

More recently Soudronic has gone even further, and devised a butt welding technique using a laser energy source. This allows even faster speeds, and enhances all the benefits previously listed for the induction weld. A continuous spiral-welding fabrication technique can be used to form tubes from steel or tinplate strip in a process analogous to that used to produce paperboard tubes and composite containers. With the improved beadless induction welding methods now available, this is being developed for metal drums and cans.

A further example of materials saving is the reduction in diameter of the can end. The technique of necking-in was developed initially with three-piece cans to eliminate the projecting beads at the top and bottom which gave rise to the phenomenon known as 'rim-riding' during transit. This not only produces unsightly scuffing and dents, but can, as a result of the high local stresses, actually introduce a weak point, and sometimes leakage.

With the advent of the two-piece can it was realised that the process could be taken further. There are two main approaches, one is the reduction by a series of small steps, and the second is a single inswept flange. The reduction (typically 16mm on the diameter of a beverage container end) also represents very worthwhile savings in metal. To improve the stiffness of thin tinplate cans of both two- and three-piece construction, the same fabrication technique is used to incorporate rolled-in hoops in the side panel.

It is quite possible to provide tinplate cans with easy-open ends in the same metal, but the manufacturing technique is very much more difficult than if these are made in aluminium. The reasons are that tinplate is harder and less ductile than aluminium, and the punching knives become quickly blunted. As they do so they can snap the remaining thickness of steel. If too much is left in situ (to reduce the chances of the first problem) then it is not easy for the consumer to open the can. A further problem is that the pre-scored cut leaves a very sharp residual edge which, due to the need for the score to be a minimum distance from the flange, projects from the can wall. A design which mitigates this problem is the Weir-open steel can end developed by US manufacturer Weirton Steel Corporation. This has a triple fold on both the body and the removable panel, so guarding the sharp-cut edges.

Both ring pull and full aperture tinplate ends are currently in use, although few of the latter incorporate the Weir-open refinement. A new form of beverage easy-open can end in tinplate, developed jointly by British Steel and Hoogovens of Holland in 1988, not only resolves the problem mentioned above, but has two other benefits. It uses less metal, and makes the tab non-detachable so reducing the littering potential. The easy open tab takes the form of two circular buttons which are pressed inwards into the can. During manufacture, these tabs are cut right through, leaving only a tiny residual link, and covered with a plastic sealant around the cut edge.

Other approaches to deterring litter have included cemented-on aluminium tabs with one edge permanently bonded to prevent removal. These are used on oil cans but do not have the physical strength to be suitable for carbonated beverages.

Aluminium

Aluminium is used to produce ranges of metal containers which are completely interchangeable with tinplate in the beverage sector, and the proportion of canned beer and soft drinks packed in each varies in different parts of the world. Figures for the European Community, produced in 1989, show the wide differences, which depend upon relative economics and in turn relate to local energy costs as follows:

Table 2.2
Beverage cans in Europe

	Steel (m)	Alu (m)	% Steel	% Alu
Italy	nil	1900	nil	100
Germany	3750	900	81	19
Austria	nil	600	nil	100
Belgium	1600	nil	100	nil
Holland	800	nil	100	nil
Spain	1650	nil	100	nil
France	550	nil	100	nil
Sweden	nil	1200	nil	100
Greece	nil	600	nil	100
UK	4698	2265	68	32
Total Europe		20,513		

Aluminium is more ductile than tinplate, so can be rolled or worked into thinner sections. As a beverage can it competes directly with tinplate but the ductility does allow a greater degree of necking-in to be carried out to reduce the diameter of the end. This has been taken further in Japan than anywhere else, with Toyo Seikan producing an eight-stage necked-in can, its diameter being reduced from the can body 65mm to 50mm. When this stage of elaboration in the fabrication is reached, the law of diminishing returns tends to apply and the materials saving may not balance the extra manufacturing costs. However, in Japan the novelty of the can's appearance gives it appeal. In normal markets, three- to four-step necking-in seems to be about the optimum. Due to the ductility of aluminium and continuing improvements in this by varying the metallic make-up of the different alloys, as well as better mechanical engineering in the forming process, a continuous programme of weight reduction has been possible. The metal content of a 33cl aluminium beverage can has fallen by 25% over the last decade although, as may be seen in Figure 2.1 below, produced by CMB Packaging, scope for further reductions is now very small.

Fig. 2.1 Changes in metal content of aluminium beverage cans, 1977-88

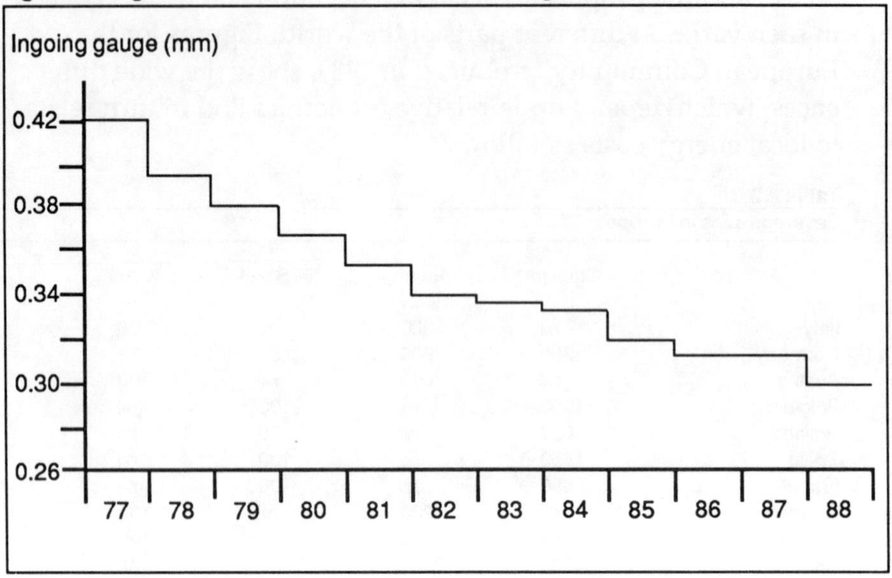

The US company Alcoa claims that 33cl beverage cans are now being produced weighing only 11.3g compared with 19.7g in 1975, a reduction of 37%. Wall thickness is a little under 10 microns. Of course no gain is totally without its drawbacks and these ultra lightweight packs need very careful handling on the filling lines as they can be unstable when empty.

Resistance welding techniques are not appropriate for aluminium but cements (adhesives) are available which can be used to combine pressed or spun components into strong containers. The nature of the material makes it easy to produce three-dimensional containers by deep drawing. Thus two-piece containers in all shapes are possible. Easy-open ends can be provided where necessary. To reduce the litter propensity of detachable ring pull tabs a number of designs are now available in which the tab remains attached to the can end. Indeed it is mandatory in about half of the US states for these to be used on beverage containers. The operation of one such design, introduced by Coca Cola in the UK in 1989, is shown in Figure 2.2.

Fig. 2.2 Coca Cola's non-detachable can opening design

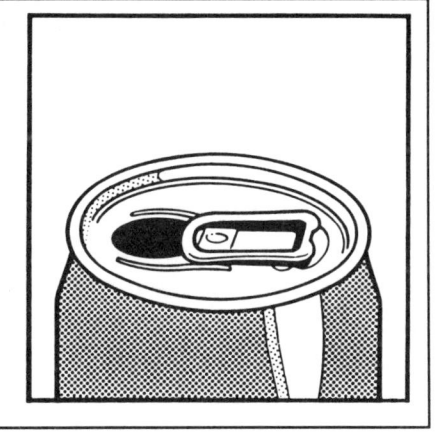

An ultra lightweight aluminium beverage can, the Clicker produced by Continental Can in the early 1980s, was formed from two components – a deep-drawn body and a pressed top which were cemented at the shoulder by a wide adhesive join. It incorporated a reclosable fold-up plastic pourer but it has not so far achieved commercial success. Other novelty packs for beverages in aluminium from Japan are made by spinning, fabrication and seaming, but these are often costly and intended for speciality or gift markets. One example of a novel pack development for a very specific product is the 'draught' Guinness can introduced in the UK in 1989. Its purpose is to provide a shock to expel dissolved carbon dioxide and nitrogen out of solution as the can is opened, and provide the characteristic 'head' for which Guinness is so famous. The secret is a small plastic chamber sealed into the base of the can before the product is filled. This has in it a tiny hole, preventing the pressure (which has reached equilibrium inside the can once it is sealed) from escaping quickly when the can is opened. The pressurised gases force a stream of tiny bubbles through the hole and achieve the desired result. The Japanese eight-stage necked-in can described earlier has a plastic clip affixed to its bottom which, on being 'pinged', sends a high frequency shock wave through the beer, achieving a similar result.

Other novelty packs include self-heating and self-cooling cans, making use of exothermic and endothermic chemical reactions inside sealed separate chambers.

So far the penetration of aluminium cans in the processed food market has been low. Flat packs of the type used for sardines offer benefits in thermal processing, since their geometry means that the processing heat has to penetrate only a short distance (typically 10-15mm) so preventing the over-cooked flavour which is sometimes associated with canned foods. Small aluminium cans may be produced from precoated and preprinted sheet materials, and a depth to diameter ratio of 1.2:1 is quite feasible. In the larger sizes the forming process damages the precoating, and a post-fabrication coating treatment is necessary. Modern electrophoretic techniques which were developed for the spraying of car bodies, allow very uniform coatings to be applied to complex shapes, so the limitations are progressively being reduced.

In the area of full aperture easy-open cans, a development from French company Ferembal solves the problem of residual sharp edges in a similar way to the Weirton method mentioned earlier, by means of an ingenious double-fold which protects the cut edge. This has been incorporated on a microwaveable plastics basin.

Pressed aluminium trays have a long history in either the rough- or smooth-walled variety. The former types have been extensively used for chilled foods in catering and institutional cooking systems, and also as containers for take-away hot meals. With the advent of the microwave oven, demand for prepared meals has grown dramatically and the market for meals prepared specifically for this kitchen appliance is growing at a similar rate.

Trays for this market are produced from coated board, plastics and aluminium foil. According to the Aluminium Foil Container Manufacturers' Association, in the UK 73% of the trays used for reheatable ready meals are made from aluminium. There has been some controversy over the suitability of these metal trays for use in microwave ovens, with the UK Microwave Association and the Electricity Council expressing reservations on the possibility of electric arcing leading to damage. This is most likely to occur if old or poorly maintained equipment is used, or if instructions are not properly followed.

There are also questions over the rate of reheating in microwave ovens. Experts have shown that heating is slower in a metal tray but more detailed trials in 1988 and 1989 indicated that there are also major variations in the heating effectiveness of different ovens, different packs and different products as well as between different areas within a single tray.

Aluminium foil can be rolled down to thicknesses as low as 5μm, but 7μm probably represents the optimum balance of material costs and processing and energy costs (below a certain point it costs more to save material than it is worth). Very thin layers, embossed to improve rigidity, are used in cigarette packs, or laminated with plastics to provide high barrier structures.

At thicknesses above 12μm, foils provide a virtually total barrier, since pinholes are rare. In lower thicknesses, typically 9- and 7μm, the inevitably present pinholes do not affect barrier performance if the foil is laminated to a thermoplastic material. The explanation is that even with a large number of minute holes, the sum total of their area is very small indeed - perhaps one-millionth of the total. The average of this proportion with the permeability of (say) LDPE, and the rest of the area being zero, is effectively still zero.

At one stage the aluminium foil manufacturers feared that much of their flexible packaging market could be taken by metallised films, but this has not happened. Certainly some substitution has taken place, but much of the metallised market is up-grading of previously used plastics.

PAPER

Paper is one of the oldest materials used in packaging: a crude form, papyrus, was originally produced in Egypt, and the modern papermaking process developed in China many hundreds of years ago. At its most simple it is made from a variety of naturally occurring forms of cellulose (mainly wood, cotton and grasses) which are macerated in water and laid out to dry under heat and pressure as a flat sheet. The water softens the outer surface of the cellulose fibres, which then fuse together when they make contact with other fibres in the suction stage of a modern papermaking machine. The bond is made stronger and hardened during the calendering stage when the paper is driven through a chain of heated metal rollers.

Papermaking

The raw materials from which paper is made are mostly derived from softwood trees, especially spruce and pine. Other tree-derived sources include birch and eucalyptus, but seed hairs, eg cotton, fibres such as flax, jute and hemp, and grasses including straw, bagasse (sugar cane residue) and bamboo, as well as leaf films, such as esparto and sisal, all have the same cellulosic structure with its ability to form strong fibre to fibre bonds, making them suitable for paper. The composition of wood pulp, after stripping off the outside bark, is about 40-45% cellulose and a further 5-10% of hemicellulose, plus a certain amount of lignin.

There are a number of physical and chemical processing routes by which this raw material is then converted into paper, all of which begin with a physical size reduction. This reduces the wood to fibres of desired length and softens their surfaces. If little or no further treatment is undertaken, a high yield, but low quality paper – mechanical grade – is the end product. Newspapers are the main destination of this, and the characteristic browning and embrittlement which takes place after these are exposed to strong sunlight is due to the lignin degrading. Mechanical paper of this type is therefore mostly suitable for short-lived products.

Most packaging papers are chemically treated after an initial mechanical breaking stage. This is a more gentle way of teasing apart the fibres and activating their surfaces to initiate bonds. The ground-up wood is cooked in a solution of either acidic sulphite (to produce bleached kraft paper) or alkaline sulphate, to produce the stronger, unbleached kraft paper. Since the lignin and much of the hemicellulose dissolve into the chemical solutions, both of these processes lead to lower yields – about 50% is the figure usually quoted.

The degree of 'beating' and chemical digestion, plus the addition of other chemicals such as sizing agents, all provide opportunities to affect the properties of the finished product. Materials as different in texture and appearance as blotting paper and dense, transparent glassine can be the end product, depending upon the process used.

At the end of the mechanical/chemical processing a thin slurry of cellulose fibres (only about 1% by weight) in water remains, and this is laid down on a moving fine mesh wire belt or in some instances on a circular drum with a fine mesh, through which the bulk of the water drains. A vacuum is then applied to compact the fibres and extract more water before the paper moves to the drying stages. Paper thickness is dictated by the number of layers of fibre which are laid down via a series of 'head boxes' through which the fibrous slurry is allowed to run.

Some paper manufacturers make use of this multiple-layer facility to produce grades having different colours or paper textures on the two faces for special effects. Another situation in which this is used is to provide a paper made mainly from recycled fibres with a thin coating of virgin fibres to improve appearance. Another is the production of the so-called 'oyster' finish in which the merest skin of bleached white fibres is formed as an overlay on un-bleached kraft to provide a mottled effect as the brown surface colour shows through the thinner areas of white.

Just one of the many highly specialised grades of paper is the type in which a thin surface layer of electrically conductive fibres, having been treated specifically to give them this property, is laid into the surface web. Such materials are used for the packaging or handling of electrostatic sensitive devices where they reduce the likelihood of static charges building up in the surface.

The most extensive use of this multilayer facility is in the manufacture of cartonboard which is really only thick paper. The definition of what constitutes paper and what is board is not universally agreed, although the most authoritative ruling should be the ISO standard which makes the transition at $250g/m^2$. Since, as mentioned earlier, the density of paper can vary, this does not exactly correlate with thickness. Other conventions based on thickness put the upper limit for paper as 250- or 300µm.

Packaging papers
Paper still represents the largest proportion of the main materials used for packaging. In the UK it currently accounts for about 40%

of the total packaging expenditure, at £2.8 billion per annum (see Table 1.2 on page 5). Its cost performance index, especially with respect to stiffness, opacity and printability, plus its versatility, and in recent years its environmental profile, have helped the material to resist the continuing substitution by plastics to a greater extent than have the other traditional packaging materials, metal and glass.

In particular paper has retained its pre-eminence by its ability to combine with all other material to make optimum use of the properties of each.

Improved techniques in the manufacture of paper, especially the use of a hybrid process – chemical/mechanical treatment – to produce what is called chemo-mechanical pulp (CMP) have raised the yield achievable, increasing the productivity of individual mills, so reducing costs, and also improving the environmental profile of the industry as a whole.

Another area in which environmental effects have been a topic of concern is bleaching by the use of chlorine-containing chemicals. This process always leads to the production of minute traces of dioxin in the water effluent, and even smaller amounts can remain in the paper pulp, the economics of washing out the last few parts per million or parts per billion simply not being justified. Although there has been not one single recorded incident of any harmful effects on humans from either source, publicity from environmental pressure groups has resulted in widespread public concern.

Dioxins are a family of over 100 chlorine-containing hydrocarbon chemicals which can occur naturally in some vegetables and are produced in any burning process if traces of chlorine are present, as well as being an incidental product of a number of industrial processes. The main route for possible contact with consumers is by ingestion if these are extracted by foods from paper-based packaging, and by skin contact, when painful forms of acne could result if high enough levels are present. Babies' disposable nappies are the main object of concern in the latter respect.

In practice neither is a real problem. The scientists involved in the paper industry have, by developing ever more sensitive detection methods, managed to identify amounts as low as parts per trillion or even parts per quadrillion, which are infinitesimally low. However, once detected at any level at all, concern can be aroused and paper mills worldwide have launched a comprehensive programme to reduce the use of chlorine in their processes.

There are two main thrusts to this, one of which is the greater use of unbleached paper. This has other benefits also since it produces a stronger material. The second is to use alternative chemicals where bleaching is essential. Oxygen and chlorine dioxide are two of the main chemical routes being adopted for this purpose. The latter significantly reduces the amount of chlorine emissions and the former does so completely.

Synthetic papers

Attempts have been made to produce wholly synthetic paper from plastics. Heat-fusing long fibres is one route and this has led to the development of Tyvek, a very strong paper-like material produced by Du Pont from HDPE filaments. Although extremely useful for some special applications, in particular medical wrappers, security envelopes and high endurance labels, it cannot compete economically at present with paper for wider applications.

Other attempts have been based on low cost polyolefin and polystyrene films loaded with inorganic pigments such as chalk, talc and titanium dioxide. These too have had some success, especially for high quality printing – computer manuals, engineers' drawings and high quality art prints, etc. A number of the situations where synthetic materials have been substituted for the previously used papers are mentioned at page 64, including Hoechst's most recent offering, VP400.

The largest Japanese producer, Oji Yuka, claims that its polypropylene-based synthetic paper is achieving record sales, 14,000 te being exported in 1988.

Probably the most significant competitor to emerge in recent years is the cavitated OPP film, such as Oppalyte from Mobil Plastics (see page 65). In many ways the physical properties of this come quite close to paper, but with the benefits of water resistance, heat sealability and an extremely low density. It is a tribute to the cost-containment improvements of the paper industry as well as the inherent economics of scale in which paper is made that even with these benefits no wholesale substitution of paper has yet occurred.

Another novel form, patented by Eastman Kodak, is based on polyester films, but incorporates microspheres of incompatible cellulose acetate. The light scattering effect at the non-fused boundaries provides it with the required opacity.

New developments

Papermaking has remained a water-based process since it was first devised. This fact makes energy a very important cost element in the economics of paper manufacture, since the water has to be progressively extracted from the initial slurry at about 1% solids to a dry paper with only 4-6% moisture content. Stacks of steam-heated rollers are used but by its nature the process is not particularly energy efficient, and for many years now research has been carried out to devise a way of producing paper by a 'dry' route, ie without the need to use suction and then evaporate off all that water. Although the industry does add some resin sizes to the pulp to promote fibre adhesion, it has not been possible so far to make this the only bonding mechanism.

Matted fibrous layers can be produced by incorporating organic resin binders and this is the process used in the manufacture of softboard, hardboard and other particle boards, but none of the processes used for these is currently capable of making a thin, white, flexible and strong material to match natural paper. The situation with respect to synthetic papers is likely to remain for the foreseeable future the same as it has been for the past couple of decades: synthetics will be more expensive and will find mainly niche applications in specific areas.

Developments in papermaking technology continue all the time and one concept which is now emerging is the so-called 'explosive steam' fibre preparation technique. A system of heating the partly refined wood pulp under pressure and then suddenly releasing this pressure promotes the separation of fibres and their surfaces rupture in an explosive way, so increasing the active surface area to make more bonding sites available. The fibres are also more cleanly separated and can be graded into size fractions for either making specific grades of paper or re-blending in precisely calculated proportions to provide optimum properties for other grades.

An alternative route with much the same end in view is one reported in 1989 by UK company DRG which specialises in high performance papers for medical uses: A range of post-fabrication treatments called the 'Barrier Conversion Process' has been developed which makes it possible to modify the pore structure of the papers in a number of ways. These may be closed up to prevent bacterial infection without affecting gas porosity, or complete barriers may be produced in situ.

Blending into paper pulp a mixture of fibres made from polyolefin plastics has been possible for some years. These can provide a strengthening matrix of long fibres within the paper, but due to their lack of affinity with water they require special treatment for any sort of bond with the natural cellulose fibres.

In 1986 Hercules Chemicals in the US, in conjunction with others, developed what is called polymer composite sheet. It perhaps has more in common with moulded pulp than true paper, since thick pliable sheets of the material are produced which can then be compressed in heated moulds to produce shaped trays, bowls and containers. The heat sets the plastics fibres to provide a rigid paper-like material.

Paper has already a good environmental image but efforts are still being put into ways of using other raw materials for its manufacture. The potential benefits are even greater if these materials can be either waste products or a positive nuisance which needs to be disposed of in any event.

Two forms of grass stalk are available in large quantities and meet the first of these criteria, ie materials for which no other economic use is possible. They are bagasse from sugar cane and straw from cereal crops. Both are capable of being used to produce very stiff board and paper materials, but even with their almost free availability the economics have not yet been satisfactory for their use as a raw material for papermaking. With an increasing demand in the UK at the current time (1990) to ban the practice of the burning of waste agricultural straw, there are pressures to find other uses, and this may encourage even greater efforts at papermaking as a means of disposal. One other raw material which can be used for the manufacture of paper and from which much had been hoped at one time (especially for the developing countries) is water hyacinth. This is a prolifically growing weed which blocks rivers and canals in the tropics. It needs to be removed and it has been demonstrated that a quite serviceable paper may be manufactured from this, but to date no large scale applications have been reported.

Finally, at the rarified end, the Japanese Research Institute for Polymers and Textiles has reported experimental results in which acetic acid bacteria, fed on sugar solutions, produce a mass of extremely thin cellulosic fibres. From these, after treatment with caustic soda, a very thin, tough 'paper' can be formed. As a production route it is of course totally uneconomic at present, but it is not unknown for whole new industries to grow from equally small beginnings.

Ranges of water-soluble paper are also available. These are produced from paper fibres to which are added a high proportion of a water-soluble polymer, usually poly(vinyl alcohol) (see page 138). Their uses include soluble packs for difficult materials such as agrochemicals, and they may also be suitable for labels on re-usable containers.

Paper can be coated with any thermoplastic material by extruding a molten layer over its surface. Low density polyethylene is the most frequently used, especially for heavy duty barrier materials, where it provides both moisture barrier and heat sealability. Other plastics used include PVDC, nitrocellulose, acrylics, TPX, and PET. This wide choice makes possible an equally wide range of materials having different performance for many applications. In 1989 the DRG Group produced the first ever paper coated with a biodegradable plastic, ICI's Biopol, a polyhydroxybutyrate/valerate copolymer. This offers good gas barrier performance and heat sealability while only marginally affecting the ability of the paper to degrade in the presence of water and bacteria.

Metallised paper, produced by either the direct route or (less commonly) by the indirect route, is also becoming more widely used, as techniques improve to overcome one of the earlier deficiencies, the evaporation of the moisture content of the paper during the process. This results in it being either unacceptably brittle, or requiring a further treatment to re-humidify before it can be used. The main applications for metallised papers are as bundling tissue for cigarette cartons where it replaces the previously used embossed foil/tissue paper laminate, and also increasingly for the production of metallic-finish labels. A major benefit claimed for metallised paper as a substitute for a foil paper structure for the label market is that moisture equilibrium can be achieved very much more quickly with a metallised paper, giving rise to less curl problems and faster, more reliable, label application on the filling lines.

Cartonboard

Cartonboard, as defined earlier, is paper in weights above 250g/m^2 or 250µm thick. The necessity to produce thick grades of paper in a multilayer process, as mentioned on page 37, makes it easy to use different grades of pulp for separate layers and, although many individual layers may be involved, most board materials are classified as solid, duplex or triplex, ie containing one, two or three distinct layers of material. Solid bleached boxboard is the top quality and, as its name suggests, it is made entirely of the purest bleached chemical pulp. Major applications are for high quality foods, cosmetics, electronic components and medical packs.

Duplex, or white lined boxboard, is a twin structure usually having a thin layer of pure bleached white paper over a thicker, partly bleached mixture of mechanical and chemical pulp. This is a very widely used form of cartonboard having an optimum mix of physical, chemical and aesthetic properties. It is suitable for almost all packaging applications, especially foodstuffs.

Triplex grades usually incorporate a core layer of lower grade, often recycled, pulp between two white or other uniformly coloured layers. These grades vary widely in quality, the proportion of each component, and in appearance. They find applications in general purpose packaging, including some foods. Another form which is more strictly a duplex although rarely called this (to avoid confusion with the type already described) is white lined chipboard. Here a board made mainly from recovered fibres is coated with a layer of white paper fibres to provide a good quality appearance and a surface capable of taking high grade printing. The main constituent is a grey pulp derived from recycled mechanical and other papers which is weak in plybond (the force holding different layers together) but very flexible. Due to the homogeneity of the re-used pulp, it usually has very uniform converting properties with easy-fold creases. These give it a consistent performance on high speed carton filling lines. Any form of non-food contact packaging (and in some instances, even for dry foods), can be manufactured from this material. Although chipboard made from recycled paper is not of a pure quality the high temperatures at which it is produced greatly reduce any bacterial contamination.

Plain chipboard, without any facing or liner, is also used for the production of fittings and base trays, also for rigid boxes or cartons. Lower weights of these – strictly paper rather than board – are also used for the production of spiral wound cylindrical drums, and as the corrugated central component (the fluting medium) in corrugated board construction.

Corrugated fibreboard

Nearly 40% of all paper packaging is used in the form of fibreboard cases, most of which are of multilayer corrugated construction. Using three layers of thin paper is an adaptation of the engineering beam principle of two flat load bearing panels separated by a rigid web. In use, one of the faces is always subject to a tensile stress at the time of any bending, and this uses the physical properties of paper in an extremely effective way. In addition to physical strength, corrugated board provides some shock absorbing properties (cushioning) and also on occasions a thermal barrier due to the trapped air space involved. Over the 150 or so years since corrugated was patented in the UK – the idea was originally for a crimped paper lining to be used in hat bands, and then for a packaging material in the USA about 20 years later – a series of standard sizes has come to be adopted in most countries.

Corrugated board is categorised in three ways: by the thickness and spacing of the individual flutings of the corrugated layer – the fluting medium; by the weight in g/m² of the facing layers – the liners; and by the quality of the paper used. The most widely used forms are known simply as A, B, C and E. Dimensions quoted in different sources vary slightly, but BS1133 Section 7 gives these as follows:

Table 2.3
Common forms of corrugated fibreboard

Fluting type	Flutes/metre	Flute height (mm)
A	105–125	4.8
B	150–185	2.4
C	120–145	3.6
E	290–320	1.2

Originally materials were either coarse flute A or fine flute B, but the intermediate grade (C) since it was developed, has now become the most commonly used type, being to some extent a compromise of the best qualities of the other two.

There is also an even finer grade than E flute called 'microflute', used as an alternative to solid board for cartons. What was claimed in 1989 to be the world's finest corrugated, Min-Dan, was produced in Japan by Crown Boards, with a flute height of only 0.6mm. At the other end of the scale there are also some 'super' grades of corrugated board with fluting up to about 80 per metre and 7mm high. None of these special grades feature significantly in world usage.

Liners to face the corrugated boards range from 125-400g/m² with 150, 200 and 300 grades as the most widely used. Today corrugated cases are increasingly used to sell their contents, so high quality printing is demanded. Three options are possible:

> *Direct printing on finished board. The uneven load-bearing properties across the flutes limit the quality which is possible by this route, and relatively simple two-colour flexo printed designs have been the norm. Great strides have, however, been made and quality has been much improved in recent years.*

Pre-printed liners. High quality flexo printed facing materials can be built into the corrugated board at the point of manufacture. Developments in paper surfaces, printing presses and the use of polymer flexo plates have made big advances possible in this area also.

Litho laminated is the lamination of printed paper or even board to complete corrugated board. Top quality printing is possible including full colour halftone designs.

Pre-printed liners, and direct printing, where this is done in sheet form, are best suited to fairly long production runs since the case design is fixed at the manufacturing stage.

In addition to the variation in the dimensions of the fluting and the weight (strength) of the facing liners, corrugated offers further versatility in the number of components which may be combined and the way in which this is done. The most widely used constructions are:

Single faced – a soft material which can be rolled up in one direction and is normally used to provide a protective cushioning function to a vulnerable item. Rolled up it can also provide a rigid cylindrical pack.

Single wall (a 'wall' is a layer of fluting material faced by two flat sheets). This is the most commonly used form. With two facing layers the board is rigid and can be printed, although not to high quality. All fluting styles from A to E and microflute may be, and are, used. This is the construction used for the majority of traditional corrugated fibreboard cases and trays. The thinnest materials, E and microflute, are also used as alternatives to carton board.

Double wall. Another very popular range, able to incorporate any combination from double A to double E, but AC and AB are among the more popular. Certain constructions are designed to provide high rigidity using an A flute component coupled with a finer grade B or even E, to provide the best surface for printing on what is to be the outside of the case.

Triple wall is the last of the main types and once again may be made from any combination of fluting grades. This category includes one of the very heaviest boards available, known as Triwall. With massive facings the material has such a high performance that cases made from this are used in place of wooden crates. They are in fact superior to traditional nailed wooden crates in some respects since the latter can fall to pieces under extremes of handling. This, plus their lighter weight, makes them used particularly for military and scientific expeditions.

Of course, not all multiwall boards are of high performance. In some countries, particularly in the developing world, there is a convention of simply counting up the number of components. Here one might be offered three-, five-, seven-, nine- or even 11-ply corrugated boards, but the materials used may be of very light weight.

Corrugated board has one drawback: it can lose much of its strength by being heavily wetted, and to overcome this it is possible to specify that the outer facing is of a wet-strength (resin treated) grade, or triple sandwich of kraft/PE/kraft paper. Starch-based adhesives are normally used in the manufacture of corrugated board, their excellent running properties being needed at the high production speeds involved. These can also be treated to improve their water resistance. Wax dipping or roller coating with wax at the flat sheet stage is also used to confer water resistance.

Although very heavy facings can be used, there are limits to the weight of paper which can be put through the heated corrugating rollers to produce the fluting material, 125–200g/m^2 being the range in normal use. To overcome this restriction some manufacturers have managed to combine two layers of fluting medium using a very strong adhesive which significantly stiffens the material at the same time, producing a very rigid structure. Unipower-S is one branded range available in the UK.

There are other variations on such paper structures. In Germany, a triple wall cross fluted material, X-Ply, has been offered by Schumacher. This was designed to overcome the difference in panel rigidity between the 'with flute' and 'cross flute' directions. Such materials are relatively easily made by sheet adhesion but this is a slow process and the benefits provided do not justify the much greater costs involved, except for very specialised uses. In any event, a well designed corrugated case is usually constructed in such a way that the most important load-bearing panels are presented with their flutes vertical to the direction of anticipated load.

The thermal barrier property mentioned earlier is exploited particularly in Japan, where packs to maintain temperatures – either cold or hot – are extremely popular. A corrugated board made from lining materials which are themselves laminated with metallised polyester film provides a reflective surface to trap heat in addition to the layer of still air in the flutes themselves.

Honeycomb structures using similar paper materials are also produced. They have a series of hexagonal cells faced with flat sheet materials and can be produced in limited sizes in thicknesses from about 10mm to 200mm. They owe more to building construction materials such as lightweight interior doors or dividing panels than packaging, although they are used to a small extent for pallet tops. Their major advantages are very high flat crush resistance and energy absorption.

Solid Fibreboard

The share of the total case market held by solid fibreboard materials has diminished as corrugated materials have improved in quality, and as handling methods (especially wider use of freight containers) have improved. Nevertheless, these materials retain about 4% of the case market by value, being used for the construction of cases and trays where conditions are very demanding.

Applications include timber framed and metal edged cases for engineering items. Here the high puncture resistance is a major consideration in its choice. Vegetable trays or cases, and trays for meat or fish account for the main sales, very often the material being coated with polyethylene to keep out moisture, the exposure to which is common to all of these uses.

Chapter 3 *Synthetic materials (plastics)*

INTRODUCTION

Plastics, although now universally accepted as an identifiable group of materials, is in fact a rather loose expression. It comes from the adjective 'plastic', used to describe materials which can be soft and malleable, allowing them to be moulded into shapes which are then fixed by heating, chemical reaction, or by cooling. As such, most materials – including certainly clay, metals and glass – can strictly be described as 'plastic' materials.

The first of what we now call plastics were discovered over 100 years ago during the search for substitutes for natural materials such as ivory, ebony and tortoiseshell. They are mainly of the type now known as thermo-set plastics, ie those which once moulded into the final form and set by heat, cannot be subsequently softened. During the 1930s to 1950s these became the major types of plastics, Bakelite being the best known name.

These materials, based on phenol formaldehyde, urea formaldehyde or melamine formaldehyde, found applications in packaging, mainly as rigid closures and fittings, their brittleness and limitations on forming processes making them unsuitable for most containers.

Thermo-plastics materials were developed during the 1930s. Celluloid (cellulose nitrate), cellulose acetate derived from natural cellulose materials, and Perspex (polymethylmethacrylate) are just a few examples, but their use in packaging was initially very restricted.

Regenerated cellulose film is also one of the oldest synthetic materials, but does not qualify as a plastic material on the grounds of either its origin or nature. It is derived from natural cellulose (wood mainly, but other sources are possible). Its manufacturing process means that it can be produced only as a thin sheet. Since it is a natural cellulose it does not soften with heat, merely chars like paper (this is the easiest way to test for the material). Indeed, its early name, transparent paper, was adopted by a major UK manufacturer as its company title.

From the 1950s on, two groups of plastics became available, distinguished by their full titles as thermosetting resins and thermoplastic resins. The first were direct descendants of the old materials such as Bakelite and its antecedents, the latter were a new form of materials for which the chemical term polymer was adopted. These are by far the dominant form of all plastics in use today. They are all based on organic chemistry (that branch in which long chains of carbon atoms can build up into giant molecules) and at present they are virtually all derived from petrochemical feedstocks – mainly crude oil. They can in fact be produced from other organic sources including vegetable materials but today's economic factors make these of marginal interest, although in the very long term they will no doubt become significant.

Most of the major polymers are derived from simple gases like ethylene and propylene (the molecules of which join together to form polyethylene and polypropylene respectively) but the technology to achieve the splitting of petroleum gases into pure monomers (as ethylene is called) and to induce these to polymerise, under the influence of pressure and catalysts, into the solid materials we need is very complex and expensive. Only the ready availability of the raw materials, and the scale on which the operation is carried out make the process economic and hence the materials so low in cost.

There are about a dozen or so plastics materials in common use, producing a spectrum of properties to match most needs. In addition there is scope for numbers of these to be combined in many different structures to provide levels of performance not available from any single material. The plastics in use today have been around for 10-40 years and meet most of the needs of packaging, but their main limitations are in barrier performance and heat tolerance.

A polymer offering improved performance in these particular respects at an acceptable cost could end the search for the 'ideal' single plastic material.

Improvements in performance or usefulness do not necessarily require the development of completely new polymers. There are many variations on a theme which can extend the range of useful applications of these materials already available, but occasionally (with increasing rarity as the number of materials already available grows) a completely new type of polymer does emerge.

Performance-enhancing developments include process modifications, combining two or more monomers (to form new classes of materials known as copolymers and terpolymers), and the blending or 'alloying' of different materials. It is also possible to modify the properties of a basic material after it has been produced – either in polymer form or in its converted form as a film or even a container. Irradiation can produce cross-linking to strengthen a material as the long molecules combine together at various points to form a strong three-dimensional matrix. Surfaces can be modified by exposure to reactive gases, and even heat or stretching can dramatically alter the properties of a plastic material. This is before we come to the almost endless permutation of combining anything from two to 10 different materials (plastics and nonplastics) in some form of multilayer structure. This approach allows the best properties of each to be employed in the most economic way.

One field where material research continues most actively is in engineering plastics where much greater performance demands exist. Research into these can sometimes lead to the development of new polymers appropriate for packaging uses, or new production methods may be devised which make some of the hitherto very expensive engineering materials available at much lower costs, and hence able to be considered for wider applications such as packaging.

Current world usage of plastics is estimated at over 70 million tonnes per annum and the distribution of this consumption with one set of estimates for growth over the next few years, is given below.

Fig. 3.1 Trends in world consumption of plastics, 1972-95

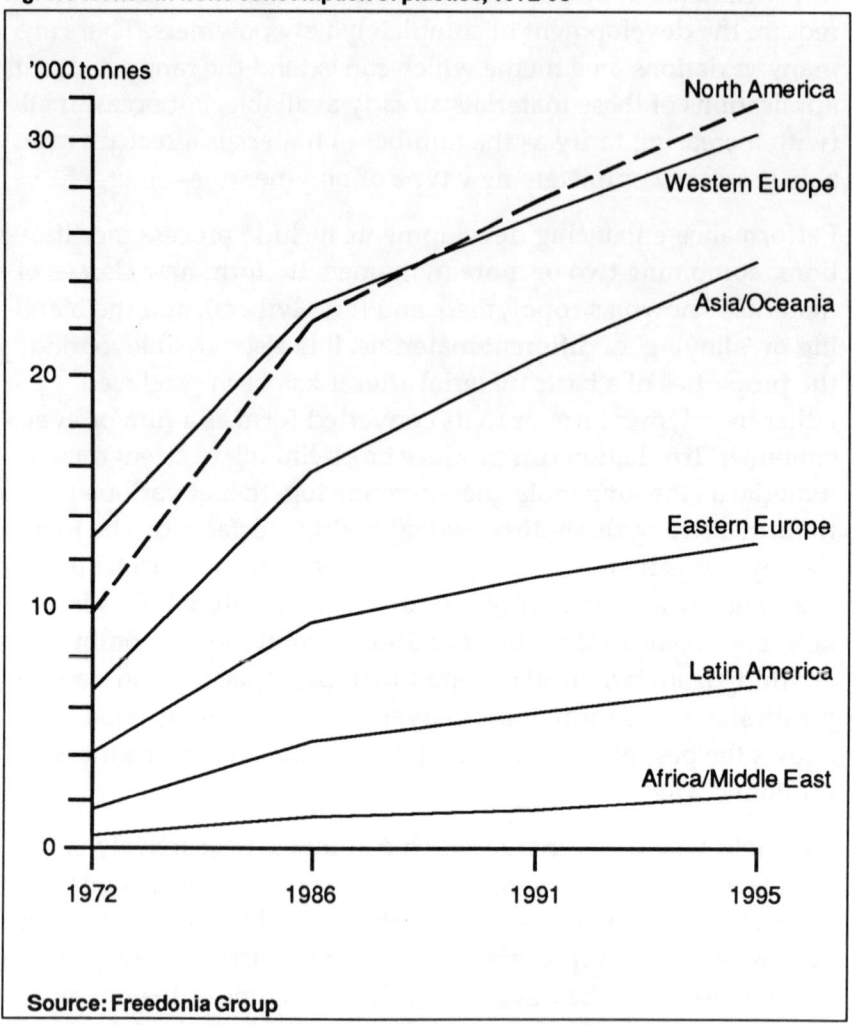

Source: Freedonia Group

Two factors are noteworthy. One is the sharp reduction in the rate of growth during 1986 as world economics entered a period of reduced prosperity. The second is that the projected rate of growth is lowest in those regions currently using the smallest amounts. This is because the units given are straight quantities. A further factor which complicates growth analysis by simple tonnage figures is that continuing improvements in performance allow steady 'down-gauging', ie more use is made of less material. Taken as percentages of their 1986 consumption patterns the growth predictions are as follows:

Table 3.1
Regional consumption of all plastics

Region	Consumption 1986 '000 te	%world total	Estimated consumption in 1992 '000 te	Total % increase 1986-92
Western Europe	23317	30.2	30500	33
North America	22594	29.2	32000	44
Asia/Oceania	16068	20.8	25000	56
Eastern Europe	9802	12.7	13000	37
Latin America	4476	5.8	7000	67
Africa/Middle East	1056	1.4	2000	100
Totals	77313	100	109500	41.6

According to the Association of Plastics Manufacturers in Europe (APME) from whom these data are obtained, packaging use accounted for 29% of the world consumption in 1986 (or 22.4 million tonnes). The split between the main families of plastics materials is estimated by APME to be

Table 3.2
Plastics consumption by type

Material	'000 te pa	%
Polyolefins	30771	39.8
Thermosets	15153	19.6
PVC	14071	18.2
Polystyrene	8118	10.5
Other thermoplastics	9200	11.9
Totals	77313	100

Each of the different materials will have its own mix of applications so it is not easy to estimate the proportions used for packaging. For example, PVC is extensively used in building components and cable covering, polyolefins mostly in packaging.

In Western Europe, the individual polymer groups are expected to experience differing growth rates. According to estimates by chemical company Exxon, these will be:

Table 3.3

Growth prospects for polymers used for packaging in Western Europe

Polymer	Production in '000te 1985	Production in '000te 1995	Average % annual growth
LDPE	2260	2270	+0.04
LLDPE	265	1025	+28.7
HDPE	1215	1630	+3.4
PP	655	1230	+8.8
PVC	1125	1100	−0.02
Styrenics	720	790	+1.0
PET	120	445	+27.1
Total	6360	8490	+3.3

Plastics offer some vitally important properties in packaging. They are lightweight, tough, water resistant, inert, hygienic, easily formed into complex or very thin sections, virtually unbreakable, highly transparent or brightly coloured, and can be reprocessed after use or incinerated to allow recovery of much of their energy content.

Although the world existed without them for millennia, there is no doubt that given the present pattern of living a vastly greater amount of other resources would be needed if they were for any reason not available. The German Plastics Manufacturers Institute (VKE), in response to a high level of criticism against plastics during the early 1980s, produced in 1986 its assessment of the impact which the abolition of all plastics packaging would have on the German economy. By allocating appropriate alternative materials for each of the plastics materials currently being used, it projected that the total consumption of materials would rise by just over 300% each year, while costs of materials would rise by over 100%. That did not take into account higher energy costs in transporting the almost invariably heavier materials needed to replace plastics. They further estimated that the volume of household waste would increase by 150%.

What is not in dispute is that the use of plastics will grow in the future. A prediction by US plastics giant Du Pont is that by the year 2000 plastics will account for 42% of the primary food packaging in the United States market – up from its 1987 level of 17%.

A brief introduction to the various forms of plastics processing will help to understand these as they are mentioned in later sections on specific polymers, applications and process developments.

All thermoplastics are melted by heat and, with one exception (rotational moulding) the first stage is a screw extruder. This works just like a mincer, polymer granules being fed into a small hopper by gravity and blended and forced through a heated zone before emerging as a continuous film or tube of molten plastic.

Fig. 3.2 Schematic diagram of extruder

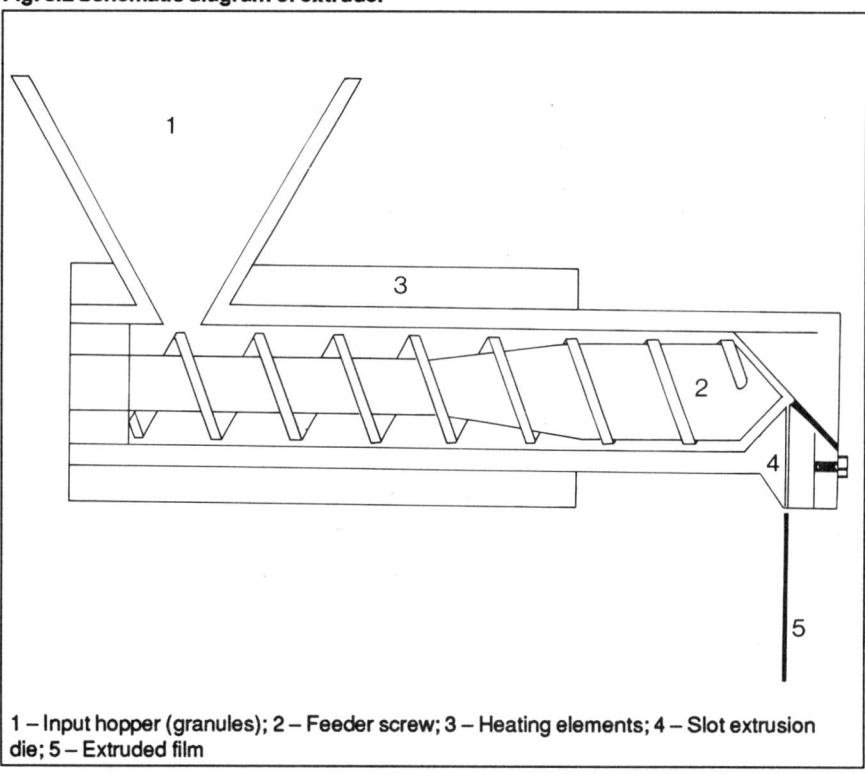

1 – Input hopper (granules); 2 – Feeder screw; 3 – Heating elements; 4 – Slot extrusion die; 5 – Extruded film

Processes which can be employed as the next stage include film casting, in which plastics are spread via a slit die on to a chilled polished roller. The polymer solidifies into a film and is then reeled up. In tubular film blowing the extrudate emerges via a circular die, producing a continuous seamless tube of hot plastic. Into this compressed air is blown to inflate the tube, the diameter expanding by a ratio of two to three times. After cooling this is then flattened and reeled-up as layflat tubing.

Fig. 3.3 Schematic diagram of film blowing equipment

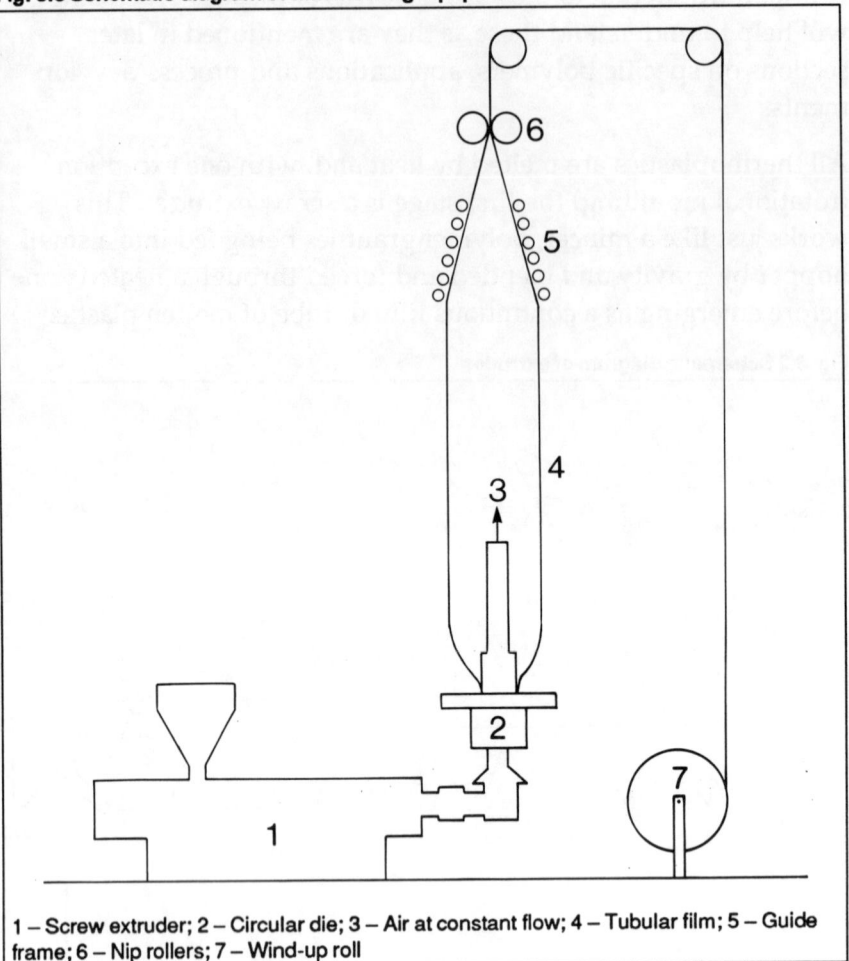

1 – Screw extruder; 2 – Circular die; 3 – Air at constant flow; 4 – Tubular film; 5 – Guide frame; 6 – Nip rollers; 7 – Wind-up roll

Bottle blowing by the extrusion-blow-moulding process involves the extrusion of a thick tubular parison which is clamped between two halves of a solid mould (pinching the base closed in the process) and air is then blown into the open top (neck finish) to produce bottles or jars. Only the outer surface exactly matches the mould and there is a limit to the precision achievable due to the restriction on air pressures which can be used.

Fig. 3.4 Schematic diagram of extrusion bottle blowing

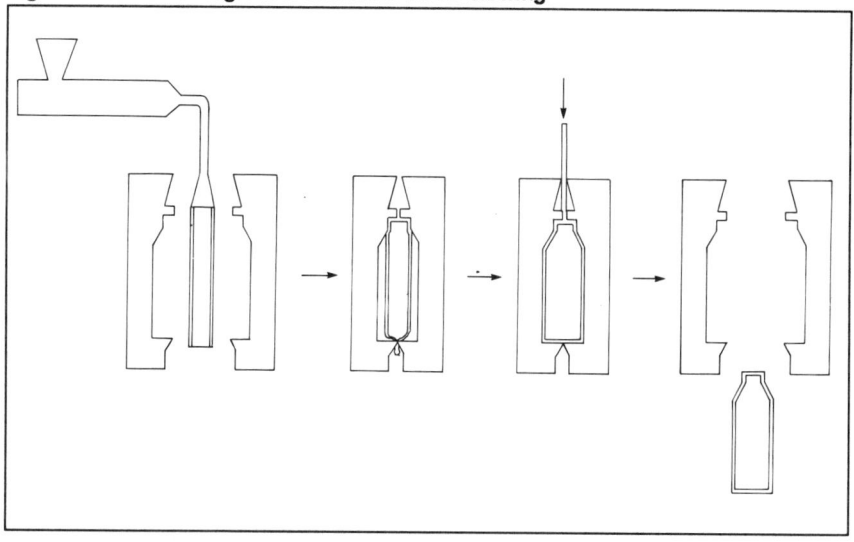

Injection moulding, as its name states, is the injection of molten plastics from a screw extruder under high pressure into a fully detailed mould. Highly precise dimensions, very fine detail and very thin sections are all possible using this technique. Injection-stretch-blow moulding is a technique in which a thick-walled preform is produced by injection moulding, reheated and then stretched by a pneumatic rod inserted via the neck. It is then blown into a second full size mould. This is the process by which all of the PET bottles used for beer and carbonated soft drinks are produced.

Thermoformed containers and fitments are produced from thick (typically 0.5-2mm) sheet plastics produced by a slot die and calendering technique. Most production is two-stage with the reeled sheet material being provided to a secondary converter who produces containers such as disposable cups, trays and fitments. The technique these converters use is to soften the sheet by heat, and force it into or over simple moulded shapes using reduced or increased air pressure. A preliminary stage called plug assist is sometimes used. In this a shaped plunger pushes sheet into a pre-shape, helping to distribute material more evenly throughout the base and wall areas, especially into the corners which invariably end up with the thinnest sections.

There are a number of hybrid systems in use. One is solid phase pressure forming in which moulded, shaped preforms are heated and stretched. Similar in many respects to this is Injecta-form, developed by UK company Clabas, in which an injection moulded preform is reheated and thermoformed into a mould to form trays. Both of these techniques allow the distribution of material to be controlled in the original moulding stage so that in their final thermoformed condition the wall thicknesses are as uniform as possible.

Another concept, pioneered by the Hercules company in the USA and developed in the UK by DRG Plastics, is called RTF, rotational thermoforming. Here a continuous sheet of molten plastics material is extruded on to a rotating series of moulds in a single continuous process. Advantages are more uniform wall thicknesses, the containers have less thermal strain built into them, and some very high output speeds are possible. The technique may be used for single layer or multiple layer structures as required.

Finally there is rotational moulding in which polymer powder is tipped into a heated metal mould which is then spun around its three axes to distribute the plastics as it melts. The material adheres to the mould wall, and fuses into a continuous layer. Quite crude moulds made of sheet metal may be used and the slow process is most suitable for larger items such as bins and intermediate bulk containers.

For any of the pack manufacturing routes involving extrusion it is possible to produce multilayer structures of two or more polymers (either basically different or varying in quality or colour, etc). The technique requires only a number of screw extruders (one for each different polymer) which feed their output via a combining die to force layers of different materials out together. Being still molten they fuse as a single material if they are compatible. If not, then a further material called a tie layer (or 'glue' in the US) is necessary. Some examples of different constructions are shown below.

Fig. 3.5 Coextrusion

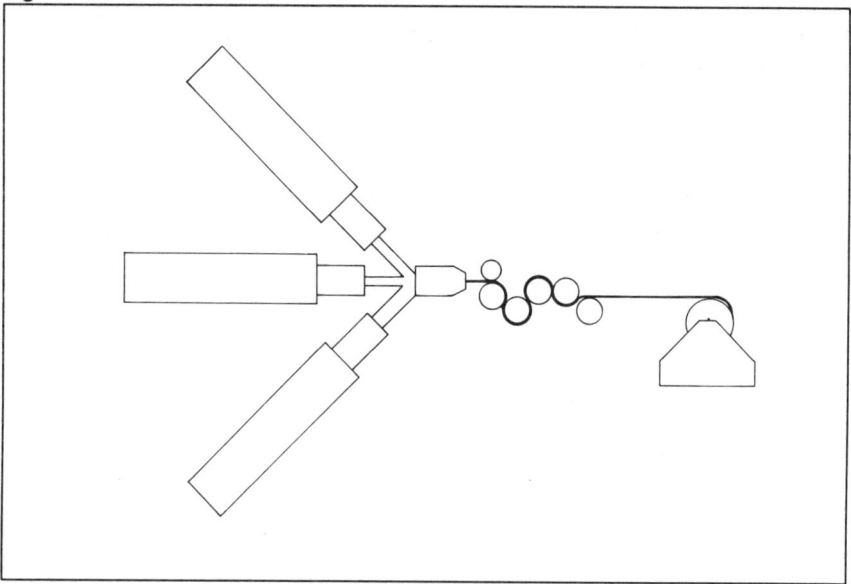

POLYOLEFINS

These may be considered as the workhorses of the plastics packaging sector because of their wide range of useful properties. As their name implies, they are formed by the polymerisation of certain unsaturated hydrocarbons known as olefins (or alkenes). Polyethylene and polypropylene are by far the most important although other members of the family, such as polybutylene and poly(methylpentene), have their own established uses.

Each is characterised by its primary building block, with each successive one in the series containing one more CH_2 group. Thus ethylene is C_2H_4, propylene is C_3H_6, and butylene is C_4H_8. The number of hydrogen atoms is always twice the number of carbon atoms. The molecular arrangement is shown in Figure 3.7 on page 68.

Polyethylene is formed of long chains of C_2H_4 units, but due to the ability of carbon atoms to form side branches, and the variations in conditions in which polymerisation takes place, the material is not always formed as molecules of standard shape, or the same size. Long straggling chains with branches is the best way of describing them; they tangle together and form a tough transparent fusible material. The three main properties of polyethylenes exploited in packaging are – toughness, heat sealability, and their barrier to water and water vapour.

In each of these properties their cost/performance ratio cannot be bettered. The fact that the atoms of carbon and hydrogen may be arranged in many different ways makes it possible to produce slight variations in the properties by using different polymerising techniques.

The first sub-range of materials developed had different densities. Low density polyethylene (the basic material) has a density range usually defined as from 0.915-0.940g/cm^3. High density polyethylene may be anything from 0.94-0.97g/cm^3. There have been attempts to subdivide the density range into more than these two, especially in the USA where four grades were specified. A medium density grade was commonly described, ranging from about 0.926-0.940g/cm^3, but today with blends and copolymers available broadening the ranges of the two main types, these intermediates are no longer considered as materials in their own right – certainly not such that they can be categorised only on their density.

Some properties vary with density, since this is to a large extent a measure of the crystallinity of the material. Properties which improve with increasing density are tensile strength, gas and water vapour barrier, rigidity and temperature stability. Properties which diminish are clarity, impact strength, elongation and heat sealability.

In recent years a number of other grades have been developed, the most significant being linear low density polyethylene. This has an almost linear molecular structure (hence its name) but it does include short branch chains.

Fig. 3.6 Low, high, and linear low density polyethylene structures

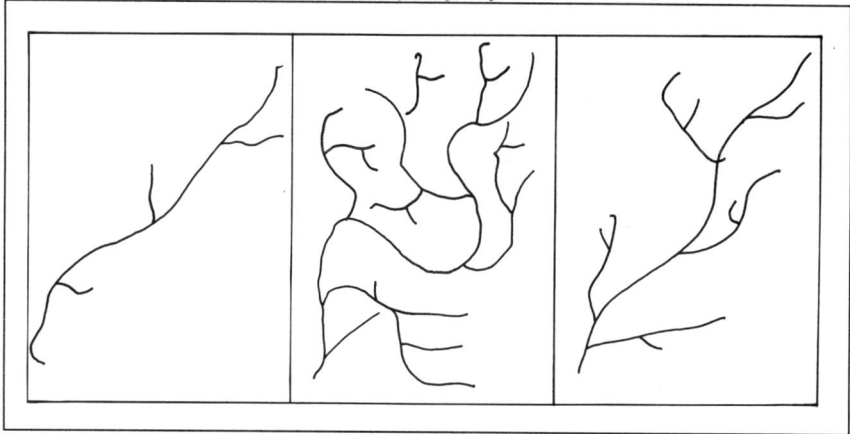

The density of linear low density resins occupies the middle of the LDPE range at 0.912-0.928g/cm^3, but their properties are in most respects superior to ordinary low density. Particular benefits of linear low density are higher physical strength in all respects, a higher temperature tolerance and better clarity. It can also be made in lower thicknesses due to its inherent strength in the molten state. It was initially offered at a higher price than low density but since the manufacturing process also offered advantages to the producers, an increased capacity became available as many of these replaced old LD plants with LLD plants and this helped to force prices down. The material was initially blended with ordinary low density as a cost-optimisation exercise, but as more has become available its use as a single material has grown. From a zero level of use in 1978 it reached 40% of total LDPE sales on the US market by 1988 and is projected to account for 70% of all polyethylene sales in the USA in 1990.

In Europe a much lower rate of penetration is quoted by supplier BP Chemicals. In 1988 only an estimated 15% of the total linear low density/low density film market was said to be linear low, but projections for a 15-20% per annum growth rate over the next five years accompanies this and this is likely to result in a catching up of the US situation.

Typical uses for LLDPE are for carrier bags, stretch films and heavy duty plastics sacks. Indeed for virtually all polyethylene situations it provides benefits.

Following the realisation of the scope for tailoring the molecular arrangements by the selection of different starting monomers, polymerising conditions and catalysts, plastics manufacturers have put much effort into refining the knowledge and opportunities this brings for all polyolefins.

Ranges of so-called very low density and ultra low density polyethylenes, having densities below the $0.915g/cm^3$ of LD, became available in 1988. Attane from Dow Chemicals at $0.912g/cm^3$, Norsoflex from Orkem, and another from Sumitomo Chemicals in Japan, are just three of the VLDPE range on offer. Union Carbide has come in with an even lower density (ULDPE) with a range of 0.900-$0.905g/cm^3$.

Physical properties for these ultra low density grades are claimed to be superior even to linear low density, with higher elongation, better puncture resistance and hot tack (making them particularly good for heat sealing through surface contamination), better optical properties and enhanced water vapour barrier performance.

Initial markets for these VL and ULDPE are based particularly on their softness, disposable nappies for babies being an example quoted by Shell Chemicals. In this market they are competing more directly with such materials as EVA and plasticised PVC rather than the traditional polyethylenes.

High density polyethylene is used as a film and, to a much greater extent, for the production of plastics containers. In the USA less than 4% of its total packaging use is for film. A major advantage of HDPE is its higher melting point, which makes it suitable for boil-in-the-bag applications. As stated, the melting point is related to density, and a table showing this relationship is given below.

Table 3.4
HDPE densities and melting points

Film density g/cm²	Melting point °C
0.92	104
0.93	112
0.94	120
0.95	128
0.96	136
0.97	145

Clarity is generally poor and heat sealing, although achievable, is more critical than for low density grades. This difference in melting point is exploited in one of the major growth areas for high density film, the liners for breakfast cereal packs. These consist of a coextruded film of low density and high density polyethylene – a heat seal is achieved by the inner surface of low density but it does not cause fusion of the outer HD layer. This allows the two surfaces to be easily peeled part for gaining access to the contents.

Pigmented grades of thin film (about 12μm) are frequently used for the production of small bags for wet foods – especially meat and fish. A particular benefit, due to the material's greater stiffness, is that they are easier to open than LDPE – 'blocking' is much less of a problem. They have also been extensively used for carrier bags in heavier weights (17.5μm thickness is typical) but there is at present a trend to substitute these by thicker (23μm) LLDPE.

A modified form of HDPE, loaded with a white pigment, has been developed by Hoechst Chemicals. Known as VP400, it has properties very similar to food grade greaseproof paper, for which it is particularly offered as a substitute. It is easy to write upon and has paper-like folding characteristics.

At Tokyo Pack in 1988 a biaxially oriented form of HDPE film (Lupic) was shown by Tonen Sekiyu Kagaku. This, claimed to be the first time its manufacture has been possible, is offered for some of the applications currently using OPP. Major improvements in clarity have been achieved and also in folding characteristics, making one of the available grades especially suitable for twist-closing for the wrapping of sweets. Another has strongly directional tearing properties, making easy-opening a special benefit.

Courtaulds in the UK also launched a mineral-filled polyethylene material 'B42' as a substitute for paper, with traditional SOS bags as the target market.

Polypropylene is often described as the most versatile all-round plastics material and is claimed to be the fastest growing. Figures for growth quoted at a conference in late 1989 were 12% pa for Western Europe, 9.5% pa for the USA and 15% pa in Japan. Its properties are basically similar to polyethylene but its melting point at 165°C exceeds any of the polyethylene grades. It is usually described as a semi-crystalline polymer, and the degree of crystallinity can be controlled by the rate of quenching and subsequent annealing during manufacture. This affects both the density (ranging from 0.895-0.920g/cm^3 – typically 0.905g/cm^3) and (in the film form) its toughness, although this effect is apparent only at very high levels of crystallinity. The high yield due to its low density, makes it particularly economic material for packaging use. The spread of packaging applications for which polypropylene is used in the UK is probably typical of other regions. These are given, together with consumption figures for 1988, below.

Table 3.5
Consumption and uses of polypropylene in the UK for packaging 1988

Form of use	Consumption in '000te	%
Film	53	35.5
Thin-walled containers	24	16.0
Boxes, crates and pails	19	12.7
Closures	26	17.3
Woven bags and sacks	12	8.0
Twines	6	4.0
Strapping	5.5	3.7
Miscellaneous	4.5	3.0
Totals	150	100

Source: RMA Statistical Review of UK Packaging, 1989

The wide spread of different packaging applications shown here bears out the claims for the material's great versatility.

As seen in Table 3.5 above, film is the largest area of use, by far the major part of which is in the biaxially oriented form (known as BOPP). It has progressively taken the market previously held by regenerated cellulose film for overwrapping and form-fill-seal applications of snack foods. By 1988 it had taken about 85% of the UK market for the two alternative materials (often referred to as the Cellopp market).

BOPP film is made by extruding a film and stretching it either by inflating a tubular bubble or, more commonly, by the so-called stenter process in which a thick extruded sheet, heated to its softening, but not its melting point, is gripped by its edges and mechanically stretched by a factor of 300-400%. The effect of this treatment is to orientate the long molecules in both the machine and cross directions, increasing the toughness and enhancing the clarity. Heat sealing is not easy for this oriented material but the facility is readily provided either by coating the film after production with a heat-sealable material (such as PVDC which at the same time provides an excellent gas barrier) or by coextruding the original film with layers of lower melting point grades on the outer faces. The latter is completely adequate for any low gas barrier applications.

Cavitated, or 'pearlised', OPP is one of the real success stories of the 1980s. This has an ultra-low density, by which it achieves the major benefit of high materials efficiency. In properties it quite closely matches paper, and has achieved some notable successes in confectionery wrapping, especially replacing glassine paper for chocolate bars wrapped at high speed on horizontal form-fill-seal equipment. Coextruded or coated grades are available, providing different barrier properties for particular applications. Another rapidly developing use is as a substrate for self adhesive label production.

Toughness, high yield and high temperature resistance are polypropylene's major assets. Two limitations have still to be overcome in order to allow it to fulfil its ultimate promise. They are a better gas barrier and, for blown bottles especially, higher optical clarity. The chemistry of the material makes the first impossible, but plenty of other techniques in which it is coated or coextruded with high barrier materials are available. In film form, PVDC and acrylic coating provide very acceptable levels of oxygen and aroma barrier.

Clarity of extrusion blow moulded bottles has been a limitation but by using newer grades developed in the late 1980s by, among others, ICI, Hoechst, Exxon and Himont, extremely good levels of clarity are now attainable. So called statistical copolymers offer the highest clarity. The injection-stretch-blow process for the production of bottles and jars – developed initially for PVC and especially PET (see description at page 57) – can also be used for producing polypropylene containers. In this the reheating and stretching of the thick walled preform has the effect of orienting the molecules in the side wall, providing enhanced toughness and clarity in the same way as these are achieved with film.

In 1988-89 a number of new ranges of wide-mouthed jars were offered for food packaging applications. These are capable of being hot filled and even post-fill sterilised. Where the materials' own modest gas barrier properties are adequate a single layer is suitable, but where higher oxygen barrier is called for it can be provided by either a central core layer of a high barrier polymer (see High Barrier Materials below) which is coinjected into the preform, or by coextrusion of the parison if extrusion blow moulded, or by surface coating with a high barrier material such as PVDC.

Stretched polypropylene filaments woven into tough fabric make another form in which the material is used. Once polymer scientists overcame the early problems of ultraviolet sensitivity for this material, it found uses in heavy duty sacks and particularly for the manufacture of flexible IBCs. Strapping tape is another area in which polypropylene is used. Yet another is as a calendered sheet for the production of translucent cartons, and for thermoforming into heat resistant trays suitable for re-heating in a microwave oven. Mineral-filled grades have been available since the 1970s, Kartothene from UK company John Waddington being the oldest. Doeflex is another, Piberplast another, and the Ferro Corporation has produced one of the latest, this one imported from the USA to Europe. All use talc or chalk as the mineral filler.

Polybutylene (or polybutene), the next in the polyolefin series, is a very tough and 'rubbery' polymer with a melting point of 125°C. It is not extensively used as a single material except for certain large flexible liquid containers. It can be blended with LDPE to enhance performance and in particular its incorporation can provide 'easy peel' heat seals.

VINYL-BASED POLYMERS

The best known of these is polyvinyl chloride, or PVC, a widely used material having many properties which make it the economic choice for packaging. It is easily processed, has high rigidity, toughness and clarity. In packaging PVC is used in two main forms. One is the rigid, unplasticised form (UPVC), the other is as a soft pliable material which contains a large proportion of a plasticising compound, usually a phthalate ester.

The main use areas of UPVC are as a clear sheet for thermoforming, the production of clear, transparent cartons, and for the extrusion blow moulding of bottles. Plasticised PVC is mainly used as the familiar cling film and for shrink wrapping, and stretch wrapping applications. The molecular structure is similar to polyethylene but with the critical difference that one hydrogen atom is substituted by chlorine, thus C_2H_3Cl is the basic unit which builds up into long chains.

Fig. 3.7

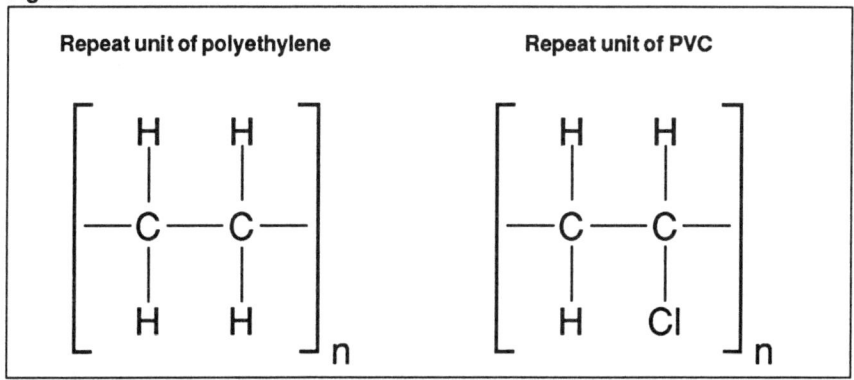

UPVC requires the use of anti-oxidant additives (stabilisers) to reduce the degree of thermal degradation which can occur during processing. The plasticised grades have been very extensively studied by food safety scientists to determine whether the amounts of the mobile plasticiser which can migrate into foods wrapped in the material are a potential health hazard. Whilst no direct likelihood of any harm arising from the use of these for food packaging has been established, an agreement was reached in 1987 with manufacturers to reduce the amount of plasticiser and also to promote the use of less mobile materials to provide a similar end product. Polymeric plasticisers are now being introduced for this reason as well as grades having lower plasticiser content.

PVC can also be injection moulded but the route has not so far been extensively used for packaging applications – partly due to the material's limited service temperature. In 1988-89 a number of chemical companies developed modifiers to increase the useful temperature tolerance of PVC. Rohm and Haas offered its Elix 300 range of additives based on glutarimide acrylic copolymer which, when used in up to 30% concentrations, can increase the usable temperature range from about 80°C to 110°C. Another additive, this time based on styrene maleic anhydride from Monsanto is claimed to provide a further vital 10C°, taking the tolerance to 120°C. This makes the material suitable for microwave oven heating, and may result in further growth applications.

PVDC (polyvinylidene chloride) – or, described more properly in the form in which it is mostly used, vinyl chloride-vinylidene chloride copolymer – is one of the best gas barrier materials available, and has been used as a coating on all forms of packaging for many years. It can be coated as an aqueous dispersion (emulsion) or applied by an organic solvent. Both methods are used, especially for the coating of regenerated cellulose and OPP film.

PVDC is also applied to the outside of plastics bottles, especially PET and PVC, to increase their gas barrier properties and make them suitable for the distribution of oxygen-sensitive liquids such as beer. The coating is usually achieved by a dipping technique, although spray coating and roller application techniques are also used.

PVDC is also used as the barrier component in a number of sandwich coextruded sheet materials. The coextrusion process allows very thin layers (just a few micrometres thick) to be built in, but even this amount provides excellent barrier. Using it in this way makes economic sense as the polymer is very expensive. Another benefit of the sandwich coextrusion is that PVDC does not come in direct contact with the metal surfaces of the extrusion die where it can cause problems of corrosion.

Although PVDC was the earliest specialist barrier polymer and now has a number of competitors, it holds its position very well, especially since, unlike certain other materials, it provides a high barrier performance against the transmission of both water vapour and oxygen.

The main rival materials in the high oxygen barrier stakes are, EVOH, amorphous nylon MXD6, and Barex (acrylic multipolymer). Competing for water vapour barrier performance are PCTFE, HDPE and PP. Coatings of these and other compounds, especially the newest vacuum metallised and vacuum applied coatings of silicon dioxide and silicon nitride, are also significant in the barrier market. Both PVC and PVDC contain chlorine which has led to a prolonged debate on their environmental acceptability, and even to attempts to ban the materials completely in some countries. The main basis of these charges is that the presence of chlorine can, under certain incineration conditions, lead to the production of hydrogen chloride gas. It can also, in other conditions relating to burning in incinerators, combine with hydrocarbon volatiles to produce a family of materials called dioxins. Certain of these are known to cause illnesses and are accepted as being undesirable. The chemicals are part of a group of over a hundred related compounds, as mentioned earlier, many of which are quite harmless but a few have been described as extremely dangerous. Incineration is one set of conditions in which these compounds may be produced and this has led to their association with the potential emissions arising from the incineration of domestic waste containing PVC packaging.

In practice only a small proportion of the total domestic refuse is incinerated in a few countries of the world. Chlorine is present in this from many other sources including food waste, and in any event research has shown that there is not a direct quantitative relationship between the amount of chlorine actually present in mixed refuse and the amount of hydrogen chloride or dioxin produced. Many other conditions influence this, including the nature of other wastes (most vegetable materials give off alkaline vapours which neutralise acidic gases), temperature, rate of burning, chimney conditions, etc. A further factor is that modern incinerators can incorporate flue gas scrubbers and are meant to operate at high temperatures which break down most of any dioxins which are formed.

One problem is that these chlorine-containing materials have had such a bad press that scientists have developed extremely sensitive test methods to detect their presence. These have reached a point where quantities as low as parts per quadrillion are measured and quoted. This equals one gramme in one thousand million tonnes – an absolutely insignificant amount, and vastly less than the amounts present in very many natural foods – beer and cabbages, for instance, according to Prof Ames, an internationally respected American specialist in cancer research.

There is a second area of concern about dioxins, which also involves chlorine, used traditionally in the bleaching process in paper manufacture. This was discussed in the Paper section.

The situation at the time of writing is that PVC is still regarded by independent observers (as well, of course, by its manufacturers) as a very versatile and cost effective form of plastic.

It is almost unchallenged in the building components area – guttering, drain pipes and fittings – on account of its durability, fire retardance and the economics of manufacture. At least 50% of the world's production is used for packaging purposes.

Much of the criticism voiced against the material does not stand up to scientific analysis but there has been an effect on the market perception and in some quarters it is being replaced by OPS, PET and other plastics on the ground that these can be advanced as more 'environment friendly'. The major manufacturers accept that this is affecting their markets to some extent, but the overall pattern of use (in all areas) has held up and shows a steady if only slight growth over recent years.

Areas of packaging growth include trays for modified atmosphere packaging, food containers (especially in the convenience food sector – prepared salads, sandwiches, cooked meats, etc), and transparent cartons, and all have helped PVC to maintain significant growth. The growing world market for natural mineral waters, most of which are currently packed in PVC bottles, has also helped fuel this growth.

Vinyl acetate, another related monomer used in its polymerised form (PVA), is a very important adhesive. Copolymers of vinyl acetate with ethylene – EVA – form an important range of plastics having very similar properties to LDPE. Their use is more extensive in the USA than in Europe and the particular properties of the material depend upon the proportions of vinyl acetate to ethylene in the final blend. These normally vary from about 5-50% of vinyl acetate. When present as a small constituent in polyethylene (below 5%) vinyl acetate can be regarded merely as an additive where it helps modify rheological and heat sealing performance.

STYRENIC PLASTICS

The styrene monomer is quite different from the olefin type, being based on an aromatic molecule (one containing the benzene ring in its structure). Polystyrene is the most frequently seen form of this material in packaging, but there are a number of other plastics which include the styrene constituent in their make up.

Polystyrene is a highly transparent non-crystalline polymer – despite its frequently used name of 'crystal polystyrene' which refers in fact to its clarity rather than its structure. It has a density of $1.05 g/cm^3$ and softens at about 95°C. In its normal form it is brittle and this has restricted its use in packaging mainly to injection moulded containers – thick-walled fancy boxes or thin-walled drinking cups. It is also extruded as a clear film but this also tends to be brittle (it has a very characteristic metallic sound when rustled) and has found only small applications in packaging. One particular use is for wrapping flowers, another is for certain fresh produce such as lettuce where the film's very high permeability to water vapour helps to restrict wilt by controlling moisture loss.

Much of the polystyrene used is in the so-called high impact form, HIPS, in which a small amount of rubber-like polybutadiene material is blended to increase its toughness, although at the expense of reducing slightly the brilliant crystal clarity. The other main use of polystyrene is as the cellular foamed material expanded polystyrene. This is discussed in more detail under Cushioning.

Recent developments have centred on improving the physical performance of polystyrene to reduce its brittleness. One significant development originating from Japan is the production of biaxially oriented grades of PS (BOPS) sheet materials which can then be thermoformed. This is emerging as a competitor for PVC and PET sheet in the USA and Japan with production approaching 100,000 tonnes per annum in each country. Production in Europe, by Klepsch in Austria and Belgian company Sidaplex, began only in 1989 but a similar market size is projected to develop there within five years.

The production process is similar to the stenter techniques described under polypropylene, and sheet can be produced in thicknesses up to 0.6mm. The material can be thermoformed at temperatures of 110-125°C – a somewhat narrower range than PVC can tolerate, and since the material has a tendency to shrink as it nears its melting point, high mechanical pressures are used in the thermoforming operation to minimise the chances of this occurring.

The material is more expensive than either PVC or PET, but with a 23% benefit in yield (due to its lower density) and a claimed higher rigidity factor, suppliers anticipate that it will be a strong competitor. Early indications have tended to confirm this expectation.

Heat resistance of polystyrene is a restricting factor. The Dainippon Inks and Chemicals Company in Japan produced a food contact grade of BOPS in 1989 for which thermal stability up to

120°C is claimed. This makes the material eminently suitable for thermoformed microwave containers, and market trials have been carried out. Dow Chemicals also has a new high temperature grade under development, for which a temperature tolerance of "up to 200°C with excellent stress crack resistance" was claimed by the company at the K89 exhibition.

On the impact performance scene a new grade of super tough polystyrene called SHIPS (super high impact polystyrene) was launched in 1989 by Stag Plastics in the UK. The company claims that fracture resistance measured by the Izod scale is 2-3 times that of traditional high impact polystyrene.

Other styrene-containing materials used for packaging include ABS, (acrylonitrilebutadiene styrene), a tough thermoformable material. Its detailed structure is a copolymer of styrene and acrylonitrile with the butadiene finely dispersed and trapped within the molecular matrix. By varying the proportions of each of the three constituents a wide range of properties can be obtained. Major use areas are for consumer durables like refrigerator door panels, but grades are also used for packaging mainly as thin thermoformed tubs or trays.

Styrene acrylonitrile, SAN, is another material which can have packaging applications. Its use is dictated by the ratio of styrene to acrylonitrile (frequently 3:1). It is offered as another choice to ABS, PVC and OPS. The gas barrier is improved by the presence of the acrylonitrile constituent but it is not claimed to be a very high barrier material since the AN is almost invariably a minority constituent.

Styrene butadiene, SB, is a copolymer which can be converted by all processing routes into containers, sheet, film, etc. It is better known by its trade name K resin and is more widely used in the USA than in Europe. It is frequently blended with other compatible resins to enhance their performance, and is particularly used for food trays and cups, and also for medical applications.

POLYESTERS

Polyesters are the fastest growing group of plastics used in packaging, finding applications as film, and strong tape, and in the production of trays, bottles and jars. Originally considered as textile or engineering plastics they burst on to the packaging scene in the late 1970s as a direct result (it is quoted by some) of the decision by Coca Cola to increase sales of soft drinks. Larger containers were the preferred way of achieving this, but problems with weight and safety of very large glass bottles made these impractical, and set off the search for alternative materials.

The high physical performance of polyesters was well known and the barrier properties, although only modest, were felt to be acceptable for short shelf life drinks. Allied to the requirement to use this material was its increasing availability, as the popularity of all-synthetic textiles waned, providing the necessary polymer availablity to meet this new market need. One further key element was the concept of the injection-stretch-blow technology for making bottles. This had been demonstrated initially with PVC, and the orientation effects, which were already known from film and tensile tapes, offered an even better performance.

The material which was identified for this project is PETP (usually shortened to PET) or polyethylene terephthalate, which is very similar to the Terylene used in textile fibres. It is inert, has high clarity and there are no restrictions on its use for food contact. Since bottles for carbonated soft drinks have to withstand high internal pressures (up to four atmospheres or 60lb/in^2) flat bases are not possible – they would blow outwards. Hence the first generation invariably had hemispherical bases which were made stable by the addition of a separate base cap. Later developments using a 'multiple dome' design withstand the internal pressure while still providing a reasonably stable base.

Loss of carbon dioxide through the walls does occur but the rate is acceptable to modern retailers and product manufacturers. Trials to reduce this by coating the outside surface with PVDC copolymer barrier resin met with mixed results – small traces of carbon dioxide permeating through the PET became blocked by the barrier layer and concentrated in the form of small bubbles or blisters on the surface. Although better coating techniques largely eliminated this problem, the need for such a coating for carbonated soft drinks quickly diminished. The widely accepted performance level in this respect, demanded by most of the soft drinks manufacturers, is that the carbonation level should not fall by more than 15% from its initial level over a 90 day period. Data published by ICI show that today this is comfortably exceeded by a 2l bottle as in Figure 3.8a below.

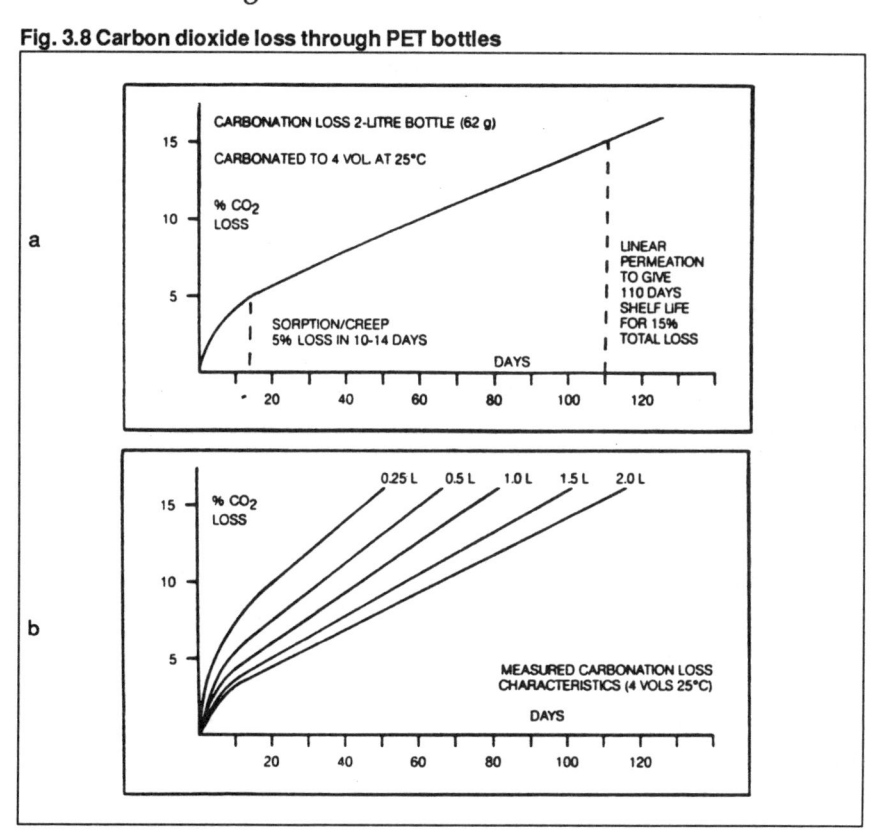

Fig. 3.8 Carbon dioxide loss through PET bottles

It should be remembered that the carbon dioxide barrier is a function of surface area so the smaller the bottle the higher the surface area to volume ratio and therefore the higher the rate of carbon dioxide loss. This is one reason (the other is cost) why the smaller bottles were slower to become adopted. This is demonstrated in Figure 3.8b above, which shows that the 1.5l bottle meets the 90 day 15% loss criteria but the 1l and certainly all smaller sizes have difficulty in doing so without coating. This figure is also by courtesy of ICI.

Table 3.6
PET bottle market share (% by volume) for carbonated soft drinks, 1987

Austria	7.7
Belgium	42.4
Eire	53.0
Finland	5.5
France	40.7
Greece	16.8
Italy	59.7
Norway	17.9
Portugal	4.0
Spain	13.8
Sweden	20.0
Switzerland	4.7
UK	38.8

Source: ITF (Pakex organiser)

The figures in Table 3.6 show wide differences, due largely to differences in legislative attitudes within the individual European countries. PETP is also used for beer bottles, but here the critical barrier need is to prevent the ingress of oxygen rather than loss of carbon dioxide. Oxygen causes beer to go 'stale'. In order to reduce this to an acceptable degree, PVDC barrier resins are almost always applied as an external coating. This improves oxygen barrier by a factor of about 100, and water vapour barrier by about 10. The size range of these containers for alcoholic beverages extends from 50ml for airline spirits miniatures to the 30l 'Beer Sphere' designed in the USA as a party pack. In Japan a large PET bottle incorporating a particularly ingenious moulded-on handle, has recently been produced by Toyo Seikan for a 5l whisky pack.

Wine, spa water (plain and carbonated), and most recently toiletries are now being sold in PET bottles. A particular range of products for which the high mechanical strength of PET is especially important is agricultural chemicals, some of which can be very hazardous in their concentrated forms. Fibrenyle of the UK – part of the Lawson Mardon Group – has developed special ranges of containers for this sector. In 1989 the world's first PET aerosol,

Petasol – also from Fibrenyle – appeared and received high acclaim for its technical innovation.

One aspect which initially limited the use of PET for the ultra taint-sensitive pure mineral waters was the inevitable presence of minute traces of acetaldehyde in the bottle wall. This has now been reduced by improved manufacturing techniques to the extent that ICI guarantees levels of only 5-12ppm in the preform, and much of this is likely to be driven off at the heated stretch-blow processing stage.

Recent research has shown that the organoleptic (taste) effects of these minute traces of acetaldehyde are heightened by the presence of carbon dioxide. Therefore the fact that carbonated spa waters, representing the most sensitive products on which to test the material, are being satisfactorily packed in this polymer, is a good indication that the problems are now resolved.

Although it is common to talk of PET, or even the more specific PETP, as if this described a single material, in practice for every polymer type the major manufacturers offer a range of formulations tailored for different applications.

A PET beverage 'can' called Petainer, developed by Swedish company PLM, has been trial marketed by Coca Cola in the USA, but not taken up for wider use due partly to environmental
still used in Iceland and for specialist

ρ in Denmark relaunched the pack in a
ₛ a clear plastic food container, calling it
an). This, claims the company, can be hot
normally lidded with a heat sealable foil
ρ-on plastic cap

ₑd jars are also in use for foods, especially
d those which are not hot filled. The heat limita-
he second processing stage which leaves the
'heat retraction memory'. If heated above about
ₛtort or shrink. The most vulnerable area for heat
the neck since this must accurately fit the intended

closure. There are two ways of stabilising this sensitive area, both in use in Japan but not to any significant degree elsewhere. The first is to subject the neck zone to a further heat treatment after the initial moulding process.

The effect is to crystallise the polymer (this is apparent by the fact that it turns an opaque white). In its crystalline form PET can then withstand temperatures up to 200°C and above. The second approach is to adopt a two-polymer sequential injection technique, in which a small quantity of a high melting point plastics is first injected into the mould – invariably from the centre of the base. This small charge of plastic is propelled along to the furthest part of the mould – the neck – by the subsequent major slug of PET. Two materials used have been polycarbonate and a special high temperature polyacrylate known as U polymer.

Grades of PET with higher temperature tolerance are being developed and hot fill temperatures to 85°C are possible in some instances. An interesting demonstration that all factors should be taken into account in assessing packaging options is that small unit portion jars of jam and preserves are already being hot filled using standard PET grades. Despite the product temperature being higher than should be tolerable, the mass of the small cold jar as a proportion of the weight of jam filled, means that sufficient heat is absorbed in heating up the jar to cool down the rest of the fill. Recognition of this practical effect and the successful marketing of small PET jars is to the credit of Skillpack of Holland, a company producing only PET containers and which has a particularly high reputation for innovation.

When improved barrier is required, the preform (see page 57) can be made by a coinjection technique and incorporate a barrier component such as EVOH or MXD6 amorphous nylon. Trial quantities of instant coffee have been marketed in Japan and a similar multilayer jar is available in the UK for this type of highly sensitive product.

Coinjected five-layer preforms of PET/tie/EVOH/tie/PET have been trial-marketed for beer in France. They are made on a specially adapted Krupp Corpoplast machine. In this form, 1l beer bottles have an oxygen transmission rate of 0.015cc/atm.day according to the manufacturer, giving a shelf life of six to nine months. A coinjection stretch blown PET/MXD6/PET bottle is also in use in the UK.

A novel bottle production technique, Corpotherm, was introduced by Krupp Corpoplast in mid-1990. Cylindrical preforms have their neck threads formed and crystallised, then the bottle is blown in the main mould and all crystallised. Hot filling up to 95°C is claimed for this.

Some very high quality, thick-walled bottles and jars made from PET are being used in Japan for cosmetics and toiletries. Many different effects – colours, prismatic 'cut glass' faceting, pearlescent and frosted finishes – are achievable. The containers are very expensive, and so far these packs have not been adopted on the same scale anywhere outside of Japan.

PET in its 'A' or amorphous form is also extruded into sheets and calendered to give high gloss material which is then used for those thermoforming applications which can bear the additional cost. Medical products are one such application. The same grades of A PET are finding applications for the production of transparent cartons. Sheet material is creased, punched and side seamed by adhesion or heat. The packs offer an attractive appearance for toiletries, textiles and small household items. In this application the A PET is competing directly with PVC – from which it is visually indistinguishable – and with sheet polypropylene, which although more cloudy is making some inroads into the transparent carton market. Offsetting to some extent PET's cost penalty against PVC, is its claimed faster cycling time in thermoforming, and the scope for using in-plant scrap more easily since there is less problem of thermal degradation, and no stabilising additives are needed.

A completely different sector of packaging, and one in which PET has experienced very high growth rates, is the thermoformed ovenable tray. This is closely identified with the growth in ownership of microwave ovens and the opportunity taken by food manufacturers – and especially major food retailers – to provide ready-prepared meals for use with these.

Paradoxically, if only microwave oven tolerance was needed, PET would not be considered since thermoformed polypropylene trays are quite capable of withstanding the 100-110°C temperatures likely to be encountered in these appliances. However, to provide customers with the maximum convenience and flexibility, food manufacturers and retailers recognise that a tray suitable for either microwave or conventional ovens is to be preferred. Here a fully crystallised form of PET is the answer. Manufacture is by a traditional thermoforming process for thick sheet but the tray remains in the forming mould for a few seconds to achieve crystallisation. As mentioned at page 79 above, the effect is visible in that the transparent tray has turned opaque white. Trays made in this way can tolerate temperatures up to about 220°C.

Even higher temperature grades are now available. Thermx, a copolyester from Eastman Kodak, has a working range for which figures as high as 280°C have been quoted. Another variant is to fill PET resin with an inorganic material such as talc or chalk. Purity Packaging in the USA announced such a material, with about 30% inorganic loading. It is said to have much improved stability at high temperatures, and to be significantly less costly than the injection moulded thermoset trays with which it competes.

In thin film form PET is always sold as a biaxially oriented material and, due to its high strength as well as having been developed in this way for its other main uses which include recording tapes, it is mostly used in low gauges, 12µm being the most common.

In this form it is often laminated to other heavier materials such as polypropylene and polyethylene for various forms of food packaging. It is also used in very high performance structures in combination with aluminium foil and high density polyethylene or polypropylene. This material, which can withstand both the thermal and physical stresses of pressure retorting, is extensively used in Japan for the so called 'retort pouch' which is a flexible pack in which food is cooked after sealing. This has never really taken off in Europe despite early market trials in the 1970s and some high expectations at that time. A similar high performance structure is, however, used for certain medical packaging applications. Unsupported PET film in a heat sealable grade is also used as a lidding film for sealed trays which have to be heated in a microwave or conventional oven.

As well as providing a high strength function to a number of multilayer structures, PET film has proved itself to be an ideal substrate for the vacuum metallisation process. When coated with a minutely thin layer – something like 400Å (or one-millionth of an inch) – of aluminium, vapourised under high vacuum conditions, the material has its oxygen barrier improved by a factor of 100-1,000 depending upon the quality of the metallising and the thickness of metal deposited. Another use of PET film, this time vacuum metallised to a much lower level (ie using less metal) is as a so called 'susceptor' film. The metallisation process and other associated treatments are discussed further at page 103.

There are, as already mentioned, other forms of polyester. One which has many of the properties of PETP, but which has been developed more as an engineering polymer for the production, for instance, of small parts for automobiles, is polybutylene terephthalate, PBTP. It is also made in film form, and has excellent injection moulding performance which have led to some specialist packaging applications.

Yet another subdivision of the versatile polyester group of plastics is PETG. The G stands for glycol and these materials are usually described as glycol-modified polyesters. The major company involved in their commercialisation is Eastman Kodak under the trade name Kodar PETG. It has been widely adopted for packaging uses, mainly in the form of extrusion-blow-moulded bottles and jars. The material is of high gloss and clarity and has very good processing characteristics. Its major application areas are for toiletries and cosmetics and foods where it competes directly with PVC. The visual appearance can often be made superior to PVC but the material is significantly more expensive. The Lawson Mardon company Fibrenyle has been among the first in the UK to offer a standard range of bottles made in this material.

The complex chemistry involved in the synthesis of the polyester group of plastics allows for the possibility of many further variations on the properties which may be available, and these in turn may offer particularly attractive benefits. The Eastman Kodak company is in the forefront in exploring the options here. One complex polymer of which it has disclosed a small amount of information is poly(ethylene naphthalene dicarboxylate), mercifully usually shortened to PEN. Trial quantities produced have shown that it is very suitable for converting into films and blow moulded containers. Tests on these have shown that the material has superior properties to PETP; specifically oxygen barrier is said to be five times as good, carbon dioxide barrier nine times as good and water vapour barrier four times as good. It is also claimed to offer increased temperature stability, making it particularly suitable for hot-fill applications. It can also be converted into a crystallisable grade by the addition of minute quantities of one of the specialist chemicals used for seeding the crystals. In this form also the properties are said to be extremely promising. At the time of writing the material has been under active development for two years but has not so far appeared in any commercial quantities.

The potential for new classes of polyester as exemplified by the PEN material is such that they could offer the possibility of being the 'ultimate' plastics, ie they are inert, require no separate additives, are fully recyclable and can either match the performance of traditional materials or at least provide an adequate level of performance for modern food distribution.

There are some other minor uses of polyester in packaging. High performance plastics strapping tape for banding pallets and bundles, is one example, although due to the inherently higher price which this commands over the more widely used polypropylene material, its use is justified only in those cases where its extreme tensile strength can be economically exploited.

POLYAMIDES (NYLON)

This group comprises a whole class of chemicals developed commercially in the 1940s – like PET, initially for textile uses.

Today about half a dozen of them have found applications in packaging although these are usually for highly specific applications where their properties merit the expense. In the USA for instance, of the 200,000 tonnes per annum used, by far the majority is for engineering components such as bearings, small gears, etc.

Polyamide chemistry is complex, and a system of naming based on the number of carbon atoms in the original monomer, which represents the size of the repeat group in the long chain polymer, has been adopted. Of these, types 6, 11 and 12 are the most widely used. A further series of sub-groups exists which are identified by a two-part number. This is based on an even more specific characterisation of the molecule. Of these the best known is nylon 6.6, but 6.9, 6.10 and 6.12 are also used. In commercial use, types 6 and 6.6 are the most important.

The particular properties of polyamide which are of greatest significance for packaging applications are its toughness and puncture resistance, grease resistance, its good gas barrier performance, and especially its resistance to flex cracking. These properties are exploited in the form of very thin films – oriented nylons – which are used as a component of high performance laminates, competing often with oriented PET (see page 81 above). The largest sources of oriented nylon films are Italy, the USA, Denmark and Japan.

Oriented nylon film, usually produced from nylon 6 polymer, can be made by a stenter process route as used for OPP and PET – see page 66. In Japan a novel manufacturing method, the so-called 'double bubble', is also used. In this a second inflation stage is carried out on a tubular extrusion line, increasing the stretch of the tubular film by a further 100% at a lower temperature.

Nylon is one of the few film materials (PET is another) which in its non-shrinkable form can be used at high temperatures. Roast-in bags and boil-in bags are made from both materials. The former concept, which is not strictly packaging, had achieved a certain popularity in the 1960s and 1970s, but has not continued to grow very much since then. Boil-in-the-bag, on the other hand, a more direct form of packaging, is a specific use of nylon which does continue. When the gas barrier properties are also being used, it is cost effective. But if these are not being exploited then the material is significantly more expensive than the alternative material HDPE, which can be used for the same purpose.

Much of the polyamide in packaging is used in the form of multilayer structures produced either by adhesive lamination, when the oriented form may be used, or as coextrusions, usually with polyethylene or polypropylene. Nylon film can be produced either as a cast or a blown film as described earlier. The properties of the final material are very much influenced by the degree of crystallinity, which in turn relates to the cooling regime adopted during the manufacturing process. Rapid quenching in the casting process produces an amorphous type of material, whereas slow cooling encourages the more regular crystalline state. Thermoformability and the degree of transparency are also influenced by this crystallinity. Since the rate of cooling cannot be so precisely controlled in the tubular blowing manufacturing process this route usually leads to the production of films with lower transparency and gloss.

On the balance of performance and economics, nylon 6 is the form most commonly used in packaging but nylon 6.6 and nylon 6.12 are popular with manufacturers blowing tubular film because of their easier processing characteristics. When coextruded with polyolefins, so long as the melt viscosities are accurately matched, the materials do not require an intermediate tie layer.

Applications for coextruded nylon/polyethylene films include bacon, cheese, meat, greasy and oily foods, coffee, gas-flushed products and, where the material is laminated to aluminium foil, this can be another variation on the retort pouch laminate.

For certain applications such as ham and special sausages the product may be cooked by the manufacturer in the pack. The film's toughness comes into its own when vacuum packaging is used, and if the product contains sharp particles its puncture resistance is also a major benefit. Vacuum packed pouches of coffee ideally demonstrate this.

Moving away from films, polyamide can be used as an outer layer in extrusion blow moulded plastics bottles to provide a highly attractive glossy surface. This may also be pigmented to give a coloured layer over a more cost effective commodity plastics such as HDPE. Bottles of this type are available for toiletries or household chemical uses.

Two special forms of nylon are also finding major applications by virtue of their gas barrier properties. The first of these, developed by the Du Pont company, represents a totally new concept in plastic material structures. The principle is to disperse a blend of amorphous nylon (Selar PA) within a much larger proportion of a polyolefin.

The mechanical agitation and melting, combined with the extrusion through the annular die, contrive to form the immiscible globules of nylon into fine platelets which are dispersed throughout the polyolefin material. The effect of these is very similar to the tiles on a roof in that they provide a series of barriers to the penetration of (in this instance) gases, forcing them to take at best a very convoluted path, see Figure 3.9 below. The particular attraction of this route is that the proportion of material may be controlled within wide ranges and it is a single (if mixed) material which means that it can be subsequently reprocessed to manufacture new containers.

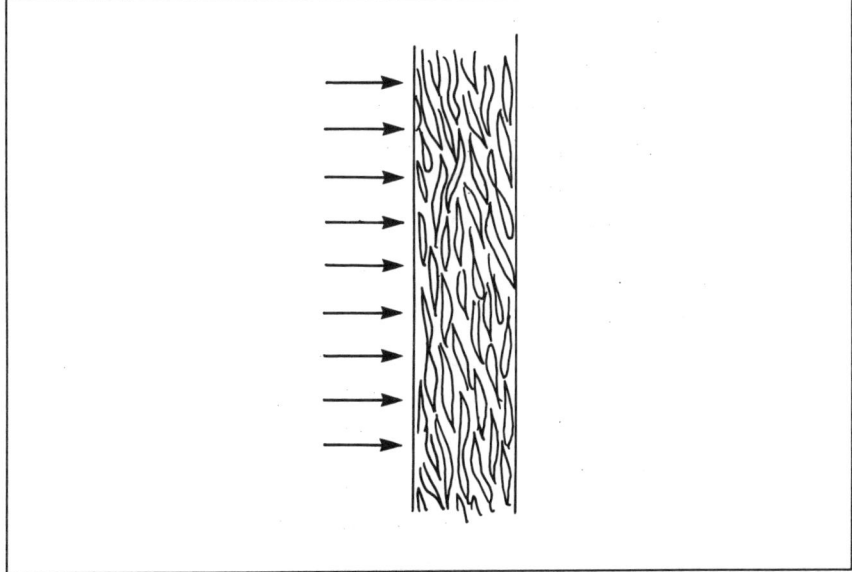

Fig. 3.9 Amorphous nylon dispersed within polyolefin

The second recent development, originating from Mitsubishi Chemicals, is MXD6, a high barrier nylon-type material (actually a semi-crystalline polyacrylamide) based on methylenediamine xylene copolymer. This is currently being used in Japan as the barrier core layer of a three-layer PET bottle for wine and beer. Polyamide is one of the few plastics materials which have a marked sensitivity to water vapour. This does, in some instances, impair its barrier properties. Selar PA, by contrast, appears from some of the data published to actually improve its oxygen barrier performance at higher relative humidities. In this respect it is exactly the opposite to EVOH, one of its rivals in the high barrier stakes.

In late 1989 at the K89 Exhibition, Dow Chemicals launched a new form of semi-crystalline polyamide for the barrier market. The initial application of this was seen as the core barrier layer in a polycarbonate/PA/polycarbonate returnable bottle for soft drinks.

GE Plastics also introduced a range of amorphous nylon compounds for gas barrier applications in packaging. Gelon A100 offers high temperature tolerance and was proposed as the barrier core in a polycarbonate/Gelon/polycarbonate plastics bottle called Eurobottle, which could be re-filled and reused in a returnable system.

HIGH BARRIER MATERIALS

Acrylic multipolymers is the name given to one range of terpolymers, the structures of which are not fully disclosed by their producing companies. XT polymer is one example; Barex, another, is much better known. This is an acrylonitrile methylacrylate terpolymer produced by Sohio Chemicals in the US initially for the production of plastics bottles for carbonated soft drinks. Like all high nitrile polymers (indeed the whole copolymer group is sometimes referred to as the HNPs), it is an excellent barrier to gases. Figures quoted suggest that it is surpassed only by PVDC and EVOH, and the latter only in dry conditions.

Barex has also excellent resistance to chemicals and organic solvents. An example of the latter is its use for the manufacture of bottles containing typists' correction fluid and its solvent. More recently, with advances in coextrusion technology and the wider range of polymers and tie layers available, new applications are confidently predicted for this material. It can be produced as thin films for laminating to other flexible packaging or as rigid sheets, again either single layer or coextrusions, for thermoforming, and it can be extrusion blow moulded into high barrier bottles. Applications in all these forms include food, chemicals and medical products.

Mobil Plastics produced a biaxially oriented film of polyacrylonitrile in 1985. Called Clearfoil it had an oxygen transmission rate of 0.6cc/m^2.day.atm and is stable at temperatures up to 200°C.

PCTFE or polychlorotrifluoroethylene, known as Aclar, is one of the few packaging materials to contain fluorine, and is generally agreed to be the best water vapour barrier polymer available. It is used when high barrier coupled with transparency is essential. Produced in the US by Allied Chemicals, it is most commonly used as a laminate with PVC for pharmaceutical blister packs. Reports of trials by the producing company quote grade 33C as having the lowest recorded value in its non-thermoformed state, 0.19g/m^2/day for a 25μm film at 38°C 95% rh. This is about one-fifth of the level recorded by a similar thickness of PVDC.

EVOH is a copolymer of ethylene and vinyl alcohol. It is the most recent addition to the army of barrier materials and was first made on a small scale in Japan in the mid 1970s. Due to its extreme moisture sensitivity arising from the vinyl alcohol constituent there were initial problems in producing it in a usable form. Today it is always used in the form of a multilayer structure protected on both sides by olefinic or other materials having good barrier to water vapour. Most multilayer films and containers are produced by a coextrusion route. Tie layers are necessary for all plastics except nylon. It can be produced as a film provided that this is then laminated or coated with barrier materials. EVOH can also be used as a barrier coating on the surface of materials, applied by spraying, dipping or roller techniques but this is not extensively used at present.

Owing to the moisture sensitivity mentioned, much debate has taken place on the effect of atmospheric water on the barrier properties. The reduction in the oxygen barrier performance of core layers of EVOH when multilayer food pouches or rigid containers are processed in high pressure steam retorts has also been an area of much research. The barrier performance of EVOH depends upon the percentage of the two monomers present; gas permeability increases, and WVTR decreases with higher proportions of ethylene. Figures published by Evalco of the USA in 1988 gave values for 25µm thick films as follows:

Table 3.7
EVOH barrier properties for different proportions of ethylene

Mol % ethylene	Oxygen transmission rate in cc/m²/day/atm. at 20°C 65% rh	CO_2 transmission rate (details as for Oxygen	WVTR in g/m²/day at 38°C 90% rh
27	0.16	0.62	105
32	0.31	0.78	59
38	0.46	1.55	33
44	1.09	3.10	22
48	1.70	4.96	22

There are now at least two grades available, having different barrier performances. The F grade has a markedly superior gas barrier. Figures (also published by Evalco) for 25µm thick films, are as follows:

Table 3.8
Permeability data EVAL grades in m²/day/atm at 0% rh

EVOH grade	Oxygen	Nitrogen	Carbon Dioxide
EVAL E	1.41	0.12	6.36
EVAL F	0.16	0.02	0.71

A further contender in the high barrier resin market, announced by Eastman Chemical in early 1988, is PEPS, a copolyesteramide. Cost considerations are likely to limit this to use as a thin core layer in coextrusions, but its emergence is yet another indication of current activity in the search for higher performance from new plastics compounds.

SPECIAL HIGH PERFORMANCE MATERIALS

There is a family of engineering polymers, some of which have minor uses in packaging. Among these is polyetherimide, a very rigid material characterised by its high temperature stability (170-180°C in continuous use, but with some specialised forms having extraordinarily high levels, even as high as 350°C). It is used for electrical components and microwave packaging. General Electric Plastics manufactures it under the name Ultem and has produced coextruded sheet materials for one-trip packaging, in particular for microwave and ovenable applications.

Polyphenylene sulphide and polyphenylene oxide (PPS and PPO) are two other high temperature polymers used mainly for consumer durable items and engineering components. Some marginal applications have been recorded in packaging, when their chemical resistance, heat resistance and mechanical strength justify their higher costs.

Chlorinated polyethylene and chlorinated polypropylene/polyethylene are highly specialised forms of the widely used packaging materials PE and PP. Their improved weather resistance and toughness make them suitable for some particularly demanding packaging applications as well as their major use in building materials.

Polycarbonate, an extremely tough engineering polymer, is best known for its use as a vandal-proof glazing, police riot shields, crash helmets and sterilisable infant feeding bottles. The main manufacturer, GE Plastics, has, for a number of years, been promoting the use of its product Lexan for packaging. Early applications were for returnable milk bottles in the USA in the late 1960s. Trippage rates of up to one hundred were claimed and at this level the initial high cost could be justified. Trials in other countries, notably in the UK in the 1970s, did not lead to wider adoption. In 1989, renewed interest in the perceived benefits of multi-trip systems for milk has led to extended trials in West Germany and Switzerland of returnable polycarbonate bottles in 1-, 2-, and 3l sizes.

Its manufacturers claim that polycarbonate is the only plastic material suitable for replacing the glass bottle for returnable milk delivery systems, since it is lightweight, extremely impact resistant, has excellent optical properties, inertness, and can be washed on the existing equipment at 70°C for many trips. Service lifetimes of over 10 years are projected, after which the polymer may be recycled. Another trial of this material for returnable milk bottles commenced in the UK in mid-1990.

Another use of polycarbonate is as one component of multilayer bottles. A combination of this with EVOH and PET in bottles produced by Danish company Holmia, used for tomato ketchup, allows hot filling or even sterilisation at temperatures which PET alone could not withstand without distortion or whitening.

GE Plastics is also engaged in evaluating multilayer polycarbonate/amorphous nylon returnable bottles to be used for carbonated soft drinks. Here they would be competing with the popular non-returnable PET for the larger size ranges. Environmental pressures offer these expensive containers the best chance they have had for some years to become adopted, according to GE. Polycarbonate is also produced as a film in large quantities for its electrical properties as an insulator, but some packaging applications such as pouches for part-cooked bakery, have been reported.

Retortable pouches is another possible area of use for which the material's toughness and high heat tolerance makes it suitable. Finally, ranges of thermoformed trays for microwave oven and hot fill use are also now available. These may be of either monolayer or multilayer constructions.

Another exotic plastics material, methylpentene copolymer, or TPX, was developed by ICI in the 1970s as a high temperature coating for paperboard, especially for use with ovenable trays. Temperature stability is up to 200°C. The material was not extensively adopted, mainly due to the difficulty in developing suitable coating technologies, and the advent of PET offering the potential to provide a similar high temperature coating at much lower cost. The patent was then passed to Japanese company Mitsubishi, which progressed it further for packaging applications. In addition to the high performance coating of board substrates, this company has since produced it as a high temperature film, and also moulded it into containers for cosmetics and toiletries. The latter exploits its particularly high clarity and sparkle. The material's density, at 0.83g/cm^3, is lower even than polypropylene, providing a high yield to marginally offset its high cost.

Polyurethanes are a whole group of thermoplastics, mostly used in packaging in the form of cellular cushioning materials. Grades are also used in the form of a coating, and for certain special uses it is produced as a very tough, thin film. A particular characteristic of the film is its soft 'feel', making it especially suitable for medical and hygiene products. Its mechanical strength and grease resistance have been utilised in certain demanding industrial and military packaging situations.

In July 1990, Rohm and Haas introduced a new copolymeric plastic material, Kamax. This is an imide-modified acrylic polymer, offering high temperature tolerance and good gas barrier. Transparent food containers and possibly returnable fruit juice bottles (hot filled) are suggested uses.

A final group of speciality plastics are those with properties in their molten or semi-molten form which can be usefully employed. Among these is a range of so-called 'graft copolymers'. These are derivatives of some of the most common polymers, modified to have grafted on to their main molecule certain other groups (or radicals). These, by their affinity to bond to otherwise incompatible materials, provide highly aggressive adhesives. EVA (see page 72) was an early form and another which, like EVA, is used as a film in its own right, as well as a coating or intermediate layer, is Du Pont's Surlyn. This is one of a family of materials known as ionomers, characterised by the presence of metallic ions in the molecule. They relate closely to LDPE and the two most important grades are based on zinc and sodium ions in the polymer. The outstanding property benefits of these over LDPE are greater toughness, oil and grease resistance, hot tack (physical strength during the molten phase) and a tenacious ability to adhere to other surfaces, especially metals.

This mix of properties makes them particularly suitable as a surface coating on paper or multi-material laminates. As a heat sealable layer they provide high strength which builds up very rapidly. This high strength is achievable even if the surface is contaminated, a particular benefit for packaging such oily food products as cheese and ham. The toughness of the material is best demonstrated by the use of Surlyn film for skin packaging, when even sharply pointed items are covered without the material being punctured.

Other graft copolymers include ethylene butylacrylate (EBA), ethylene methylacrylate (EMA), ethylene acrylic acid (EAA) and ethylene methacrylic acid (EMAA). All can be produced in film form but are most cost effectively used as blends with LDPE, as intermediate layers in coextrusions or as high quality heat sealable layers.

One final group of materials which are still in the realm of research but which could revolutionise plastics is the so-called LCPs, or liquid crystal polymers.

First used for ultra-high performance fibres (Kevlar is one), their potential wider applications in film and moulded items have been the subject of research since the early 1980s. There are two main types, one of which, lyotropic, can be produced only directly from a solution, to be spun into fibres or just possibly into film via a slit die. The other group, called thermotropic, has greater potential interest. These materials are characterised by a very sharp melting point and in their highly liquid form the molecules to some extent align themselves in the direction of flow, so providing directional rigidity. Some people have called them self-reinforcing plastics for this reason. By controlling the processing conditions, the directionality of the long-chain molecules can be influenced and it is even possible that by using electro-magnetic fields quite specific molecular alignments may be possible to provide the ultimate strength from the materials. Xydar is an example of this type of LCP, which has been used in the USA for a microwaveable dish, but one penalty for its high temperature performance is that it needs to be processed at nearly 500°C.

A newer material, for which processing temperatures under 300°C are expected, Vectra from Hoechst-Celanese, is based on a random copolymer of hydroxybenzoic and hydroxynapthoic acids. High cost still limits the use of such materials but, as pointed out in other sections, new techniques are discovered which have a habit of becoming used more widely and finally find applications in the commodity areas.

These materials have also another property used in digital display for calculators – they become opaque when excited by an electrical impulse. Even this exotic property has been suggested for exploitation in a special packaging application by Ajinomoto Foods and C. Itoh in Japan, with their ACT film, a PET/LCP/PET laminate. A clear window can be made opaque when a tiny voltage is applied. The falling cost of disposable batteries and the Japanese fascination with novel forms of packaging explain this conception.

MISCELLANEOUS PLASTICS

In addition to the major plastics used in packaging, there are other ranges which offer some special properties and hence have only limited application. They are usually more expensive than those which are more broadly used. Among these at any one time will be found the engineering materials which offer high performance but which have not yet made it into the high level of production which traditionally brings down the cost. These materials can also include some of the most exciting potential major use materials as and when demand and production factors combine to make their production more economic.

In some instances the fortunes of the material may be waning rather than waxing but remain in use either in specific countries lacking new technologies or because they have certain unusual characteristics.

Rubber hydrochloride is among the earliest of the plastics materials and, as its name suggests, it is not derived from petrochemicals. It is a very elastic material with reasonable water vapour barrier performance but is a very poor gas barrier. In Japan in recent years the material has been rediscovered and even in the US there has been some mention of its possible revival owing to the factor already mentioned – its non-dependence on petrochemicals as a feedstock. For the present, a lack of any unique properties and its high chlorine content make it unlikely to see any significant short term return to use.

Chapter 4 *Flexibles and composite materials*

REGENERATED CELLULOSE AND ITS DERIVATIVES

It is 80 years since a commercially successful method of continuously manufacturing regenerated cellulose film (RCF) was devised, and 60 years since the technology for coating with a moisture protective layer was developed. For 40-50 years the material enjoyed a steady growth in packaging applications but once oriented polypropylene film became available its properties so closely matched RCF in many of its critical areas that it began to be adopted as a lower cost substitute. So closely are the two materials identified today that economic commentators frequently refer to them jointly as comprising the Cellopp market. The rate of substitution has slowed down markedly and in most regions the division has steadied at something like 80% OPP to 20% RCF.

Regenerated cellulose is a polymeric material since it comprises long chain molecules of repeat units, but it is not a thermoplastic and is neither produced via a melt phase, nor capable of being shaped subsequently by heat. The material is produced from high purity wood pulp (eucalyptus is especially suitable) by dissolving the cellulose fibres in carbon disulphide, then adding sodium hydroxide which converts the solution into viscose, the main component of RCF. From this point, after a few days – the so-called 'ripening period' – the gelatinous material is removed through a narrow slit orifice on to a casting drum on which it then passes through further stabilising liquids, mainly sulphuric acid. This regenerates the film by coagulating the viscose solution. After passing through various washing baths the material is plasticised to make it less brittle – and hence usable as a packaging material – by the addition of ethylene glycol or propylene glycol. Only the latter material is allowed for this purpose in the USA.

At this raw stage the material is extremely moisture sensitive and for the vast majority of applications it is then coated, using either nitrocellulose or PVDC barrier/heat seal lacquers, the latter providing much the superior performance.

Production figures quoted by different sources vary, but the overall ratio of RCF to OPP in Europe is generally thought to be about 1:6 on a weight basis, and perhaps 1:10 if calculated on area, since RCF at 1.5g/m^3 is considerably denser than OPP at 0.905g/m^3.

The situation differs between individual countries depending on their traditional market and patterns of packaging. Since there are only a few producers worldwide an agreed convention for naming the grades was adopted many years ago. With some minor deviations these are as follows:

P	*plain, uncoated film*
MS	*moisture-proof nitrocellulose coated*
MXDT	*one side PVDC coated*
MXXT	*two side PVDC coated.*

The last-named of these is the most important and the highest performing grade. It is subdivided further by reference to the method used for coating the PVDC layer. When the suffix 'A' appears, the treatment has been by an aqueous dispersion coating, and if 'S' is used this denotes a solvent dispersion route. The A grade has slightly better barrier performance since the aqueous treatment allows some of the residual strains to be removed from the RCF film during the process. Barrier performance for the main grades is approximately as follows.

Table 4.1
Barrier performance of RCF grades

Code	Type	WVTR in g/m².day at 38°C 90% rh	Oxygen transmission rate in cc/m².day atm at 50% rh	at 95% rh
P	Uncoated non-seal	1700	45-75	800-1200
DM	NC coated heat seal	8	30-45	150-400
MS	NC coated non seal	6	30-45	150-400
MXXT/S	PVDC coated heat seal	7	8	8

Unlike most other flexible packaging films, regenerated cellulose is not usually designated by thickness but by 'gauge' or $g/10m^2$ over a range 280-600. The conversion table for these gauges is as follows:

Table 4.2
Thickness of RCF gauges

Gauge $g/10m^2$	Thickness µm
280	19.8
306	21.6
320	22.9
335	23.6
340	23.9
350	24.9
360	25.4
391	27.4
440	31.0
445	31.2
460	32.3
500	35.3
600	42.7

As can be seen from the above table, the method of manufacture allows some very precise gauge variations to be produced. In practice not all manufacturers would offer some of these very close subdivisions.

Although RCF has lost many of its traditional overwrapping and snack packaging markets to OPP it can still provide excellent performance when used as a laminate. As seen from Table 4.1, all grades of RCF have good barrier performance to oxygen when tested in dry conditions, and this is greatly enhanced when the material is coated with PVDC. The combination of this high gas barrier coated material, with LDPE, with its good water vapour barrier performance, provides a tough multi-barrier structure with a number of applications.

The decline in RCF's share of the Cellopp market is believed to have reached its low point and if not expected to begin increasing its share, RCF is at least believed now to have a stable future ahead. The producing companies in the US, Europe and Japan have reduced their capacity by closing plants and have upgraded those kept in commission. The effect of this is that costs have been contained and measures to protect the environment from the manufacturing effluents have been implemented.

RCF is a versatile material; it can be dyed in a range of attractive colours and is also an excellent substrate for the metallisation process (see page 103) Combining these two techniques produces some dazzling visual effects.

It is pertinent to mention here the position of RCF with respect to the question of 'environmental friendliness'. Being produced from natural renewable resources is seen by some as a major benefit, especially in the long term. Resulting from this natural origin is the fact that the material is biodegradable in its uncoated form. Neither fact provides a totally clean bill of health since high quality wood is needed for the production and in its most frequently used coated forms the material takes some time to degrade if left on the ground as litter.

With regard to the use of pure pulp as the source of raw cellulose, it has been demonstrated that RCF film can be made from other cellulosic feed materials including straw, but this is a much more difficult and hence expensive route. If the material could be produced economically from recovered waste paper sources, it might well have an excellent opportunity to lay claim to being one of the most 'green' of packaging materials.

Cellulose derivatives

Cellulose acetate is one of the derivatives of cellulose mentioned as being a very early form of plastic. It has some excellent properties including very high transparency and gloss and is easily converted into cartons, fitments and window patches by thermoforming, adhesives and solvent welding.

Further derivatives of this, cellulose acetate propionate and cellulose acetate butyrate, have generally similar properties but greatly improved toughness.

In one of the major areas of application, the production of transparent cartons, high clarity calendered PVC has taken much of the market especially since a number of proprietary creasing techniques have been developed to overcome that material's previous deficiency, a tendency to whiten and fracture at the creases. This in turn is under commercial pressure from APET sheet and polypropylene.

The use of CA as thin films for window patches in cartons and envelopes has also come under threat from oriented polystyrene film. The material has a high density and its manufacturing route is more complex than for the normal thermoplastics which are made in much greater volume. Although the CAP and CAB materials are made mainly from renewable resources and they do not contain chlorine, the manufacturing process is expensive and also requires the similar stringent controls needed for all cellulose processing to minimise any unacceptable emissions.

MODIFIED AND BLENDED PLASTICS

Mention has already been made of the wide scope for modifying the properties of plastics. Four main approaches are possible: combining different polymers in multilayer structures; blending or mixing dissimilar materials; incorporating specific additives to promote the desired effects; and modifying just the surface layer of a plastic material or pack.

Multilayer structure and laminates

By whichever route the various multi-material structures are formed, they are providing enhanced performance in an economical way, and the market is now showing signs of high growth. Estimates of the growth rates over the next four years published by the Business Communications Company in the USA, based on a report in December 1988, were as follows:

Table 4.3
Projected consumption of multi-component plastics films, 000 tonnes.

Sector	1988	1993	Average growth rate in %
Food packaging	771	961	4.5
Non-food packaging	218	254	3.1
Non packaging	190	227	3.5
Total	1179	1422	4.1

From the same source estimates of the individual types of film likely to be involved are provided.

Table 4.4
Estimated growth rates of the likely fastest growing polymers

Material	1988	1993	Average growth rate in %
OPP and coextruded PP	145	227	9.3
Polyester	29	45	8.9
Oriented nylon	8	11	6.8

Combining discrete layers can be achieved in three main ways: coextrusion, extrusion-lamination and adhesive lamination. The number of permutations therefore possible is vast, just taking into account the dozen or so commonly used thermoplastics. If, in addition, other thin materials such as regenerated cellulose film, paper and aluminium foil are considered – as is possible using the second and third processing routes mentioned above – then even more variations may be considered.

Most of the interest in high performance plastics packaging over the past 10-15 years has been in the provision of improved oxygen barrier, recognised as a most critical performance-limiting area. The other main restricting factor, a lack of heat tolerance, can be resolved in a number of ways, including the use of materials such as polypropylene for in-pack sterilising. CPET or polycarbonate can be used if microwave or traditional oven heating is required. Alternatively the use of a cool aseptic filling technique makes it possible to use most plastics. There are also developments with high temperature styrenes and PVCs as well as new grades of polyester and copolyesters. The main barrier materials have already been described under High Barrier Materials as EVOH, PVDC and amorphous nylon. Incorporating these with other materials by the coextrusion technique may require the use of an intermediate tie layer dependent upon the degree of affinity between the separate layers. Unfortunately the most widely used plastics in packaging, the polyolefins, have particularly poor affinity with any other thermoplastics materials, polyamides being the exception.

Laminated flexible films are often made from two or three materials with adhesive bonding layers. Most use polyolefins for the heat sealable layer, EVOH or PVDC as the barrier component and, if necessary, a high strength material such as PET or oriented nylon.

Coextrusion may require tie layers and with the three main materials this gives five layers, a very commonly used arrangement. If the two outermost layers are of the same material then only three extruders are needed, since this and the tie layers can each be split into two streams as in Figure 3.7 on page 60.

Increasingly it is becoming the practice to speak of six or seven layers, the extra layer(s) being of 'regrind' material, normally positioned between the barrier constituent and the outermost layer. Many companies, such as Carnaud Metalbox, Cobelplast, ONO and Wolff Walsrode, produce semi-rigid multilayer laminated or coextruded materials from which thermoformed containers having high barrier performance can be produced. These may be used either in the form of pre-made containers or as reelstock for thermoform-fill-seal equipment.

A high barrier structure produced by Wolff Walsrode in 1989, PAXX is a nylon/EVOH coextrusion, with gas transmission rates of $0.2 cc/m^2.day.atm$ for nitrogen, 1.0 for oxygen and 1.5 for carbon dioxide. These figures are an order of magnitude lower than the company's PVDC-based material Combithen PX.

One unusual form of multilayer structure used for thermoforming is that produced for the French Erca aseptic form-fill-seal system. It relies upon a deliberately weakly-bonded surface layer, usually of polypropylene, which is stripped off under a sterile environment to expose a sterile surface. This system is extensively used for aseptic packaging of dairy and other food products.

Multilayers are also employed in bottles, jars and special 'melt-to-pack' systems such as DRG's RTF (rotary thermoform) trays. One variant being produced by this technique has a seven-layer configuration: PP/regrind/adh/EVOH/adh/regrind/PP.

The solid-phase pressure formed (SPPF) containers being produced by Shell Chemicals' Rampart Packaging, and the Hitek packs produced by Reed Packaging using the cuspation dilation technique are other examples.

In the USA, American National Can – now part of Pechiney Cebal – incorporated into its retortable Omni-can multilayer pack a desiccant to help keep the EVOH barrier component dry. This was blended into the tie layer which was used to adhere the

EVOH to the outer polypropylene layers. This pack is produced as a coinjected preform and stretch-blown in a second stage to produce parallel sided 'cans' capable of being filled, sealed and retorted on existing metal can filling lines.

When considering the design of multilayer structures it is possible to make use of different proportions of materials to arrive at the most cost-effective solution. One set of information helping in such a decision, based on data compiled by Reed Plastics Packaging is as follows:

Table 4.5
Properties of main polymer types
All values based on 25μm thick films

Polymer	Density g/cm^3	Oxygen TR in cc/m^2.d.atm at 23°C	WVTR in g/m^2.d.at 38°C 90% rh	Price index by weight (MDPE = 1.0)
MDPE	0.95	2300	5	1.0
PP	0.90	2300	8	1.2
OPVC	1.36	90	10	1.3
PS	1.05	5400	90	1.4
OPET	1.24	80	15	2.2
Nylon 6	1.0	40	340	2.4
PVDC	1.7	2	1.6	3-4
PAN	1.15	12	75	4.8
EVAL F	1.2	0.5	60	9-10

SURFACE COATINGS

Surface coatings may be provided by traditional methods – lacquers and varnishes, by roller application, extrusion from hot melt, or by immersion – and the best known of these are PVDC, nitrocellulose and, more recently, acrylic lacquers. EVOH, a high barrier material, has been applied in this way but may itself require a further coating of a water protecting material. Although in principle the use of multiple coatings can provide extremely high performance, the economics and the potential technical difficulties of the multiple adhesion mechanics make it an uncommon practice. Just one of these relatively rare materials, recently announced by US manufacturer W R Grace, comprises a triple layer of PVDC, EVOH and nylon.

Wax impregnation is another form of coating used mostly for paper and board materials. In the hot dipping technique normally used, wax is absorbed into the surface pores of the substrate. Dip coating of rigid containers such as PET with PVDC to improve the oxygen barrier has been mentioned. It may be carried out either on the small preforms or on the stretch-blown finished container. In the former case the stretch characteristics of the two materials (preform and coating) must be very similar, and the adhesion, which is subject to considerable stresses, must be very strong. Fortunately modern materials are capable of coping with this. A similar requirement applies to printing on metal sheets which are subsequently pressed and shaped into three dimensional containers. Coatings on glass, mainly for decoration, light barrier or physical protection, are discussed at page 22.

In recent years the most significant new coating technique to emerge has been vacuum metallisation. In fact the principle was observed (and patented) by Edison in the 19th Century, and was used on a small scale for the production of hot stamping foils and for the decoration of special fitments for many years. The 1970s saw its expansion using large scale production plant to metallise thin films.

At first the process was used mainly to provide decorative effects, and coated regenerated cellulose film was the main material treated. The improvement in oxygen barrier performance was fairly marginal since MXXT is itself excellent in this respect. A very significant element of the total cost of metallising is the proportion of time needed to load the sealed chamber, pump down to the very low vacuum conditions needed, and start the process, in comparison with the actual time taken to run the reel of material past the metallising head. The length of runs can therefore be optimised by using the thinnest possible films, and 12µm PET provides this benefit. It has been established that the oxygen barrier performance in particular can be greatly improved by this metallic coating. Vacuum metallised PET, cellulose, oriented nylon and polypropylene are now established as materials in their own right.

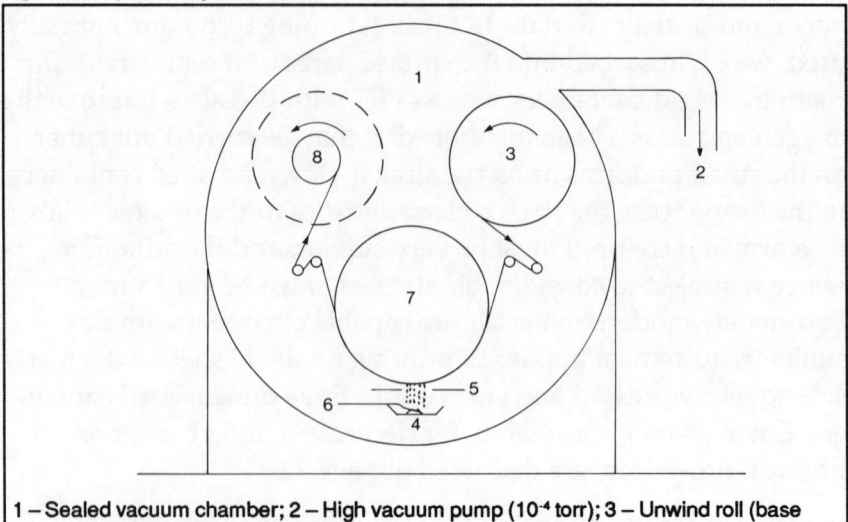

Fig. 4.1 Simplified layout of a vacuum metalliser

1 – Sealed vacuum chamber; 2 – High vacuum pump (10^{-4} torr); 3 – Unwind roll (base film); 4 – Electrically-heated boat crucible, containing feed metal; 5 – Heat shield; 6 – metal wire feed; 7 – Chilled drum; 8 – Take-up roll (metallised film)

Figure 4.1 shows a schematic view of a metallising chamber. Some of these are now 5m in diameter and can metallise webs 2.2m wide. An alternative to the direct metallisation process is transfer metallisation. This, which is really a large scale derivative of the hot foil stamping process, was developed to coat very thick materials and paper. These pose difficulties due to short run lengths for the former, and moisture boiling off under the high vacuum conditions for the latter.

Release-coated carrier webs of thin films (PET or OPP) are metallised and coated with a lacquer. The metallic layer is then transferred to the intended substrate via a heated nip roller. Van Leer's Valvaco operation in the Netherlands has most experience of this technique.

Barrier performance of metallised materials varies with the weight of aluminium deposited and its degree of compaction. Factors which affect this are the temperature, distance from the boat crucible to the film, level of vacuum, temperature of crucible, and cooling effectiveness of the chilled drum, as well as the nature of the film surface. Much effort has been put into establishing some method of defining the barrier quality of these materials using weight of aluminium per unit area, electrical conductivity in ohms per square, and optical density. The last of these is the quickest, simplest and is non-destructive, hence it is often used. Due to the high levels of opacity, a log scale is used, the so-called OD units. Good barrier materials have normally values over 2.0 and can be measured up to 4 or 5 OD units. As mentioned, because of the many variables the system does not given an absolute correlation, but as Figures 4.2 (a) and (b) below show, there is a very clear overall relationship between the barrier performance and optical density.

Fig. 4.2 Relationship between optical density and barrier performance

Optical density (transmission)	Transmission %	Transparency	Opacity
0	100	1.0	0
1.00	10	0.1	10
2.00	1	0.01	100
3.00	0.1	0.001	1000
4.00	0.01	0.0001	10 000

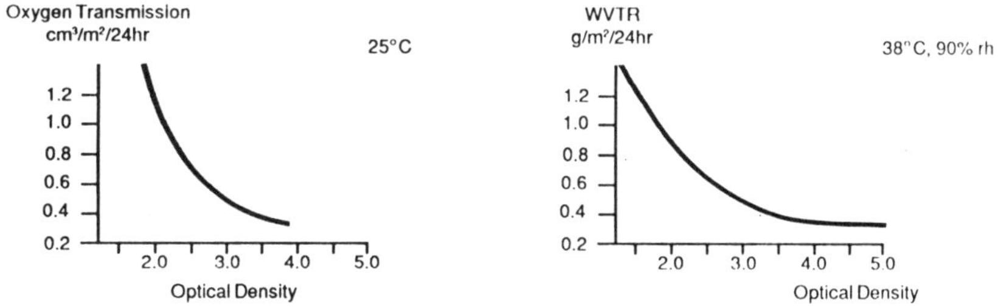

Oxygen Transmission cm^3/m^2/24hr — 25°C — vs Optical Density (2.0–5.0)

WVTR g/m^2/24hr — 38°C, 90% rh — vs Optical Density (2.0–5.0)

It has long been possible in laboratory scale experiments to also coat flat materials with extremely thin layers of inorganic compounds such as oxides. Work by Du Pont reported in the 1970s referred to barrier enhancement as a possible application of this, but its programme was mainly on inorganic phosphates prepared in situ from a solution.

More recently, work has been reported on the direct vapour deposition of silicon oxides and silicon nitrides by evaporation under high vacuum conditions, in a technique very similar to vacuum metallisation. Trial materials have been available in

Japan since the mid-1980s and commercial samples since 1988. The first commercial application of materials, produced by Toppan Printing, were used by Ajinomoto for a range of spaghetti sauces. Another Japanese company, Toyo Inks, has patented the use of a 100nm thick layer of silicon dioxide as a barrier on a polypropylene film, as a component of a paper laminate for liquid cartons. Mitsubishi describes a two-side silicon dioxide coated film of PVOH as another barrier structure. Yet another Japanese company, Unitika, has patented a coating with aluminium nitride. These few examples of published references confirm the current high level of research activity into the subject.

Silicon oxide coating requires even higher temperatures than can be achieved with resistance-heated crucibles used to vapourise aluminium metal, and the development which has made these possible is the high energy electron beam. This provides an extremely concentrated form of energy which vapourises only the small zone of the crucible fill on which the 2,000°C beam is concentrated. Film thicknesses of 50nm (one-twentieth of a micrometre), if they are continuous, can provide an excellent barrier while retaining the flexibility of the thin PET substrate. Figures for oxygen transmission rate quoted by manufacturer Toppan for a silicon dioxide coated PET/CPP laminate are 1cc/m^2.day as produced, and 2.5cc/m^2.day after retorting at 120°C for 30 minutes. WVTR is 0.6g/m^2.day tropical. The comparable figures for PET/EVOH/PP laminates are 1.0, 20.0 and 4.5 respectively.

Exposure to a combination of high humidity and temperature can reduce the barrier performance of metallised films if the metal is not adequately protected. Silicon oxide coated materials do not face this danger. Since post-fill retorting of flexible pouches is very big business in Japan, this can be an obvious advantage.

Silicon oxide coating produces a distinctly yellowish tinge to the laminate and trial materials produced in Japan by Toyo Inks and Toyo Seikan using silicon nitride instead of oxide give materials with similar barriers but having a water-white appearance.

The electron beam coating process is claimed to be much more energy effective than the earlier techniques, and this allows the cost of silicon dioxide coating, according to D Chadroudi of Suntek Inc in 1988, to be brought to about the same level as existing metallised materials – 10 cents per square metre – from its development levels which are currently some three times this.

Other techniques are also being researched in the quest for the best barrier and lowest cost route. Low temperature arc vapour deposition (LTAVD) is one such technique being offered by Vapour Technologies of New York. It is claimed to be extremely versatile and to make possible the deposition from a vapour of thin layers of almost any metal or ceramic material on a wide range of substrates. At the 1990 Interpack show, the Eastman Chemical Company in conjunction with Airco Coating Technology launched QLF, a PET bottle coating system based on a related technique.

The IKV Plastics Institute in Aachen is working on plasma polymerisation, the production of a thin film in situ by exposure to gaseous monomers followed by a microwave plasma, which initiates the polymerisation of a thin film on the surface.

Other coatings used in packaging include release finishes, which are mainly needed for self adhesive labels, see Label Materials. Another is latex cold seal adhesive. These materials have made possible major improvements in machine speed, since the small but finite time needed for heat to pass through the thickness of the substrate and fuse the normal sealable layer is no longer required. Furthermore, heat sensitive materials which could not readily be used are now also suitable for wider applications on form-fill-seal machinery.

Most cold seal coatings are based on latex with stabilisers and bond improvers. They provide either non-peeling adhesive joints or, if desired, peel cleanly by the combined layers of adhesive stripping from one of the substrate surfaces. Particular care is needed in the storage of these materials since they are affected by heat and exposure to air.

A novel 'non heat sealing' technique offered for use in blister packs was announced by Mosheer of Switzerland in 1989. It involves the use of a dry ultraviolet-activated adhesive on the backing card and/or blister edges. Brought into contact under pressure, all that is needed to initiate adhesion is a brief exposure to ultraviolet light.

Another coating, from French company Cebal, is used to seal plastics films to metal packs. The sealing layer is an epoxy base, blended with polypropylene. The epoxy adheres strongly to the metal, and the polypropylene to the film, together forming a very strong bond.

Additives

Many materials, eg plasticisers, are added to plastics in quantities from ppm (trace amounts) to high percent levels. Members of the whole group are often called processing aids. Their purpose is always to improve the performance of the host material by providing an additional property or by suppressing an undesirable tendency. Anti-oxidants are important in PVC; anti-blocking additives are needed for grades of LDPE, as are additives to promote the opposite performance when stretch wrapping materials are being produced.

Anti-static additives are often used in polyethylene film to reduce the tendency for web materials to stick to the forming collars of form-fill-seal machinery and also to make it easier to open pre-made bags. With the explosive growth of the electronics industry this particular form of additive has assumed greater significance, as the handling of most plastics readily induces electrostatic charges. Some of these are in the kilovolt range (although not dangerous to humans due to the very low current). They can, however, easily ruin a sensitive microchip, and with it an even more expensive item of electronic equipment.

Various forms of treatment are used. Some involve the incorporation of carbon to make the film partly conductive, so leaking away the charge from the surface as it builds up. Others use humidity-attracting additives to produce a thin surface film of moisture to achieve the same effect. Yet more researchers are now working on making certain polymers inherently electrically conductive. Polyacetylene is one of the most promising in this field.

Barriers to ultraviolet light are also often required for the packaging of light-sensitive products. In the past many of these have involved the addition of a yellowish coloured compound to filter out the specific harmful wavelength. Recent developments from

Japanese company Sumitomo suggest that finely divided zinc oxide particles (30-50nm) achieve the same effect. Mitsubishi proposes traces (100ppm) of a naphthalene dicarboxylic acid derivative for the same purpose.

Tiny thin flakes of mica, a naturally occurring transparent mineral material, are blended into one of the grades of Du Pont's barrier resins, Selar OH Plus. A six-fold increase in barrier performance is claimed to be achieved by a 30% addition of mica to the EVOH barrier. Its mechanism is similar to that described for the incompatible Selar RB resin. Discrete platelets are aligned by the flow of the plastics during processing into a kind of interlay of small barrier tiles. Shell Chemicals has also patented the use of from 10-35% of mica in a terpolymer for use in multilayer containers.

Metallic particles and flakes have also been added to plastics being used for the production of microwaveable trays and containers, with the intention of promoting high local heat. The phenomenon of metallic particles being excited by microwave irradiation is well known. Most of the attempts to exploit this have been in the form of extremely thin coatings of aluminium on plastics (usually PET) film by a vapour deposition technique (see metallisation at page 103).

The coated film is usually adhered to a thicker board material for ease of handling. Placed under a pizza base, for instance, this susceptor pad, as most are called, promotes high temperatures to give a crisp dry base. Temperatures at the surface of this susceptor material may reach as high as 200°C but the mass involved is, of course, very low. A variant of this is the use of vacuum-deposited stainless steel. It is reported that this metal oxidises at about 200°C, so providing an automatic cut-off to prevent overheating.

One area of packaging which is almost exclusively the province of the Japanese is that of the so-called 'freshness agents'. These are chemicals placed inside sealed packs in the form of small porous sachets where they reactively modify the atmosphere (see page 124). Related to these are plastic film materials offered in Japan which contain finely divided silica and other inorganic materials as an odour absorber. Film materials claiming this property are sold to consumers for use in refrigerators as well as being used by food manufacturers. At least one company produces a range of thermoformed trays made from a silica-filled polyolefin for which they claim a ripeness retarding performance.

Metalbox (now CMB Packaging) has developed an oxidisable polymer Oxbar as a component of a multilayer food pack with the same purpose. It absorbs oxygen, extending the shelf life of the food. The oxidation is catalysed by an additional metallic compound ingredient in the plastic.

One more recent spate of claims from Japan which appears to be related to these silica additives is the so-called 'far infrared active' packs. These are claimed to prolong the shelf life of a wide range of fresh foods from meat and fish to vegetables, by "converting far infrared irradiation". There is no agreement, even in Japan, on the actual mechanism involved, and some published laboratory results have been challenged. Such beneficial results as have been recorded are based on empirical observations and may be due to a mechanism not yet identified.

Blends and alloys

The distinction between these classifications is not so clear in the plastics world as it is with metals, but this is certainly one of the most interesting areas of current development in plastics. US specialist consultancy Kossoff and Associates predicted in 1988 that growth for these materials will average 9-10%pa over the 10 year period 1986-96 (reaching 1.45 million tonnes from a base of 0.58 million tonnes). Highest growth is predicted for Japan and the materials expected to feature most commonly are PET, polyarylate and polycarbonate. High barrier materials is one specific application area of direct packaging interest.

Some of these involve the use of compatibilising agents which act as dispersed adhesives. Others depend on a natural affinity between materials, while yet more act on the simple mechanical trapping of particles of one material in a matrix of another. Examples of recently publicised products will give an indication of just some of the current trends. Selar RB (already mentioned as a barrier polymer at page 86) is blended with the completely incompatible HDPE, producing small dispersed platelets – see Figure 4.3. The effectiveness of the barrier achieved increases as more of the Selar amorphous nylon is incorporated but there is a fall off in improvement at between 15% and 20% addition as seen in the graph below published by Du Pont.

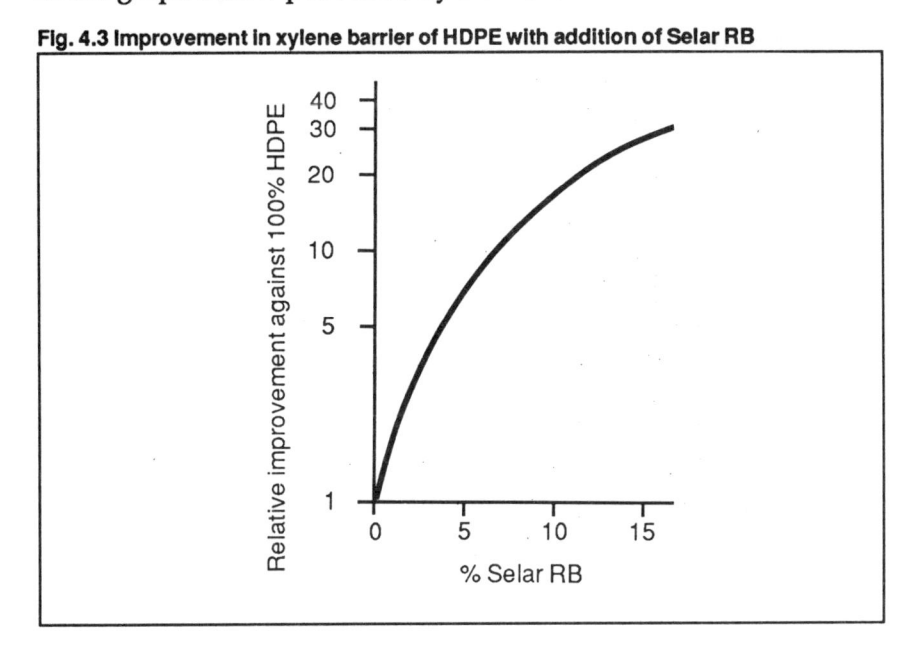

Fig. 4.3 Improvement in xylene barrier of HDPE with addition of Selar RB

G E Plastics has patented a blend of amorphous nylon and polycarbonate which offers potentially good gas barrier performance. The same company has also developed blends of polycarbonate and ABS.

Monsanto describes an example of 'third component' compatibilising resin routes in its blend of ABS and nylon. Properties are said to be superior to those of either of the two main constituent polymers.

BASF in Germany claims also to have made the first commercial production of a polyethylene/polystyrene blend by using a modified compounding process and an unspecified compatibilising ingredient. The material, introduced at K89 in late 1989, is offered as a cheap alternative for certain applications currently using PVC – in particular thermoformed packs for dairy produce.

Properties are enhanced, processing can be on any normal equipment, and one particular benefit with this as with all similar materials is that coextrusion of the blend can be achieved with other polymers compatible with just one of its major constituents.

The Malaysian Rubber Producers Research Association is working on a very tough alloy of natural rubber and polypropylene.

Mentioned here because it involves all-plastics, although more properly considered as an additive, is the development by the Osaka Industrial Technology Institute of a method to enhance the ink or paint receptivity of polyolefin surfaces, which is one of their limiting features in packaging. A blend of stearyl methacrylate copolymer introduced into the main polyolefin bleeds to the surface where it remains anchored, providing a keying layer to bind print or coatings.

These are just a few of the many blends and alloys now under development or commercially available.

Surface treatments

One final category of polymer modification is based on chemical treatment of the molecules in the accessible surface layer. Best known of these is fluoridation, which converts the surface polyolefin molecules into a range of fluorine-containing compounds of the family exemplified by PTFE (Teflon). This is one of the most inert materials known, and the presence of even a thin surface layer of such materials reduces dramatically the scope for solvent to penetrate into the wall and then permeate through it.

The technique is relatively simple: fluorine gas is mixed with nitrogen to provide the blowing air on the extrusion blow moulding machine, treating the whole of the inside surface at the same time. Airopak from Air Products has been available for a number of years. Weight loss by permeation of solvents through the wall of the plastic container is reduced by varying degrees dependent on the nature of the solvent involved. Some figures produced by Air Products are given in Table 4.6 below.

Table 4.6
Permeation test data for hydrocarbon-based solvents packages in HDPE

Solvent	Control container % weight loss	Airopack container % weight loss	Relative barrier improvement
Carbon tetrachloride	28.26	0.05	565
Pentane	98.10	0.21	467
Hexane	61.29	0.19	323
Heptane	24.26	0.08	303
Xylene	42.52	0.21	203
Iso-octane	4.54	0.03	151
Cyclohexane	22.34	0.15	149
Toluene	61.90	0.52	119
p-Xylene	59.20	0.54	110
1,3,5-Trimethylbenzene	15.85	0.18	88
Mesitylene	15.83	0.18	88
Trichloroethylene	5.70	0.30	19
Benzene	36.68	3.65	10
Chlorobenzene	32.05	5.41	6
1,2-Dichloroethane	11.55	2.89	4

Performance data for two particular solvents are also shown graphically in Figure 4.4. Liquid DIY products and agrochemicals are major use areas. Others offering a high potential include paints.

Fig. 4.4 Comparison of solvent barrier, Airopack vs HDPE

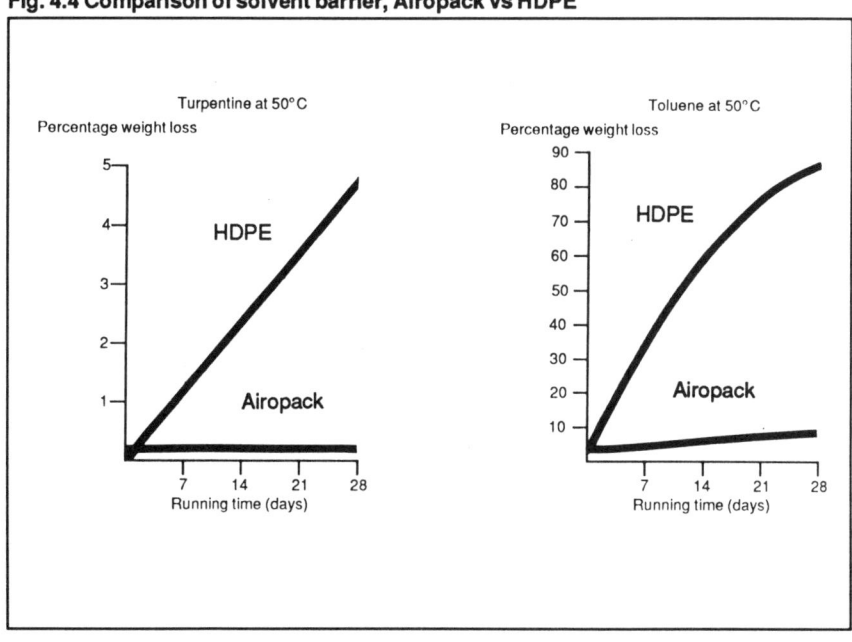

A quite different purpose is served by a fluoridation treatment devised by German company Lohmann, details of which were published in 1988. Continuous treatment of the surface of thin plastics films at the point of manufacture is proposed as a more efficient way of providing a wettable surface to promote ink adhesion. Not only is the initial surface tension improvement (the normally used method of measuring this property) better, but it is maintained over a much longer period.

Another form of surface treatment to improve barrier performance is currently being developed in a co-operative venture between Eastman Chemicals and the Airco Coating Division of the British Oxygen Company. This is based on the application of a plasma-enhanced chemical coating, but few details have yet been disclosed. It may relate to the work at IKV, see page 107.

Sulphonation, the treatment of the inside of a container using a mixture of sulphur trioxide gas and an inert gas, has been developed by Dow Chemical for use specifically with HDPE containers. It produces a similar organic solvent barrier layer on the surface and its main applications to date have been for fuel containers, and for fuel tanks on vehicles.

An alternative method used by Dow Chemicals for the same purpose is to blend in solid sulphur trioxide on an inert carrier into the surface layer of the plastic container as it is blown. The sulphur trioxide gas is released by radio frequency radiation, when the intimacy of contact, and precise location ensure that the sulphonation reaction is most effective.

COMPOSITE STRUCTURES

Although this review is on individual packaging materials, the number of necessary cross references indicates that there are many areas in which packaging materials or systems derive their performance by the combination of two or more different material types. Flexible films is one instance, paper, metal, regenerated cellulose and plastics are all traditional forms and now, as mentioned at page 106, they are even joined by glass in a rather special form.

There used to be a very specific type of packaging container known as a 'composite'. The term was coined to describe a spirally wound cartonboard tube, on to each end of which was seamed a tinplate closure. Today, with the much greater use of multi-material structures as base materials, and their subsequent combination into even more permutations to form packs, the expression is becoming used in a much wider sense.

Deriving directly from the flexible sector as a material, but owing much to the carton industry in the constructional form, is the range of liquid-tight paperboard cartons. All of these are based on paper which represents the major constituent accounting for 80-90% of the mass. All types are equally dependent upon at least one thermoplastic layer to provide both sealability and liquid tightness. Where very high performance is required these also incorporate a layer of aluminium foil. The reference from Japan about the opportunity to incorporate a thin layer of the silicon oxide 'glass' into these packs (page 106) would make microwave heating of processed foods in their packs an option. Three companies dominate this sector – Tetra Pak from Sweden (easily the largest), Norwegian Elopak and Swiss-owned PKL Combibloc – but there are others including three or four producers in Japan, the International Paper Company of the USA, the Swiss SIG company and Bosch from Germany.

The exact constructions used vary between individual companies, partly in respect to the product for which they are used. Although there have been some non-food applications – motor oil was one – the overwhelming majority of use of these cartons is for liquid foods and in some instances, semi-liquids or pastes.

Tetra Pak, which will not consider any non-food use at present, produces both the machinery and the materials for use with it. The company dominates the aseptic packaging market which in turn accounts for the largest part of the company's business.

Materials are produced for use on these machines in many different countries of the world but all of them are made to strict Tetra Pak specifications and almost all are supplied in reel form. When intended for use on an aseptic packaging machine, the construction is commonly PE/unbleached kraft/adhesive/aluminium foil/PE/PE. The two layers of LDPE – coextruded – are selected so that the layer in contact with the foil has the highest possible adhesion, and the surface in contact with the food has the least possible taint propensity.

Unbleached kraft is used because it is the most cost effective in respect to mechanical strength. Bleaching not only adds to cost but marginally weakens the material, making it necessary to use a slightly higher grade for equal performance. A white finish is obtained on these materials by a clay coating on the surface. Environmental benefits can also be claimed, since chlorine usage in the paper mills is greatly reduced. Combibloc and Elopak differ from Tetra Pak mainly in that the cartons they manufacture are supplied as pre-cut blanks – the first is side sealed, and the second is of a different style, being gable top.

In an independent assessment of the environmental impact of all forms of liquid food containers, published by INCPEN in November 1989, the cartonboard 'brick pack' recorded the lowest total energy requirement of all packaging forms. This was true whether the basis used was per container or per litre of liquid food (see page 11).

Liquid carton systems offer the facility for packs to be produced in sizes ranging from about 200ml to 3l, although not all suppliers offer the full range.

Tetra Pak has the widest range of styles from a gable top to brick-shaped, plus two very novel and relatively recently developed systems. One of these, Tetra Top, is the first ever to employ an on-machine injection moulding system to mould a plastic end with its built-in easy open and reclosable pour tab. The other, the 'D' shaped Tetra King, is made from a sandwich structure of polystyrene/foamed polystyrene, and hence strictly does not come within this discussion. Equalling these in novelty and also typifying the multi-material approach is the Hypa-S system from Robert Bosch of Germany. This is produced from a paper/plastics/foil/plastics reel stock and is fitted with semi-rigid aluminium ends.

Another important group of composite containers also produced from plastics and paper are thermoformed plastics trays with board panels to provide stiffening. Best known of these is the Tritello from Akerlund and Rausing. It consists of a pre-cut and scored board blank inserted into a mould with a thermoformed sheet of plastic formed on to this, fusing to a heat-activated coating. The necessary rim to allow the fitting of a separate lid is also formed at the same time from the plastic sheet. Other systems which are variations on the theme include Gemella, Danapack, Gemini from CMB and Sandherr. Differing from these mainly in degree are the thin-walled thermoformed pots with an in-mould paper label. This paper component provides most of the wall's stiffness.

A particularly imaginative composite container devised 10-15 years ago under the Airfix name consists of a precut and creased laminated board blank which is inserted into an injection mould. A thermoplastic material is injected, which forms a series of ribs providing both a means of locking the panels together and also where necessary a stiffening cage. The system was licensed in a number of countries and acquired a number of names including Kepak and Duopac. Although still produced in the UK by DRG Plastics it is most widely used in Japan where Dainippon Printing produces it under the name Pillard pack.

Step Can from CMB is a stretch extruded clear PET tubular pack fitted with tinplate, seamed-on ends, which has been used for processed fruit and vegetables. A proprietary heat stabilisation technique allows this to be heat processed after filling. Production ceased in 1990 on economic grounds.

Letpack in its relaunched (1989) form is the latest arrival from Akerlund and Rausing. This is made as a continuous extruded polypropylene tube with which are combined a series of continuous foil/paper labels to provide a high gas barrier. Plastics ends with easy open tabs are subsequently sealed on to cut lengths. This pack can be retorted in a steam/pressure retort. With all these (and many other) novel packs making good use of combinations of different materials, it might be thought that the traditional spiral wound composite has no place. Far from it; these are still used in great numbers, and modern equipment and printing techniques make possible better quality decoration than ever before. The inside surface can also be both impermeable and liquid tight. Nor is it essential to use metal ends; thermoformed or injection moulded plastics caps and plugs provide many design opportunities.

From this selection it is possible to see just some of the versatility which combining two or more materials offers. In terms of optimum use of materials resources this type of design can frequently be the most effective.

Chapter 5 *Ancillary materials*

ADHESIVES

Modern adhesives have come a long way from the animal glue and simple starch based types first used in packaging. Most in use today are synthetic and all are produced to formulations derived from scientific testing. Modified starches (dextrins) dominate the traditional water based ranges for sealing paper based materials, particularly for the manufacture and conversion of corrugated board. Also water based, the synthetic PVA emulsions can now be formulated for widely different substrates including totally impermeable surfaces such as glass. Their range of usefulness has been extended as newer ingredients have become available.

Solvent based adhesives are still extensively used for coating self adhesive materials, although on both cost and environmental grounds there are moves to replace these by either water based acrylic materials or 100% solid adhesives applied as hot melt.

Hot melts themselves are a well defined group of adhesives which are increasingly being used in high speed packaging applications where their instant bond and lack of solvents meet the performance requirements. The ability of manufacturers to widen the scope for application of hot melt adhesives owes as much to developments in application systems as to a better understanding of the rheological and formulation characteristics of these materials.

For the very highest performance the single- or two-part polyurethane adhesives are still used. Their major area of use is in laminating flexible materials which may have to withstand boiling water and physical abuse. Structures include the retort pouch and some high performance medical packaging systems.

In many laminating lines it is now practice to carry out extrusion laminating. The technique used here is to extrude a layer of molten polymer (usually LDPE) into the nip between rollers carrying two other webs. In this way not only is solventless adhesion achieved but the polyethylene layer can constitute a further barrier component in its own right.

Cold seal ranges based on latex have been described at page 107. Cyanoacrylate adhesives – the so-called superglues – find only very special use in packaging, mostly in the assembly of small components in fancy closures or fitments for medical or toiletry dispensers.

ADHESIVE TAPES

The traditional paper based tapes with a water-activated adhesive, for so long the mainstay of the case closing market, have gradually been ousted by self adhesive plastics based ranges. Although the paper tapes could provide excellent fibre tearing seals, their effectiveness was, and is, very dependent on the care with which they are applied. Surfaces which are too absorptive can dry out the moisture rapidly, leading to a weak bond. Their bond strength also falls off if they are allowed to become too dry before application. Pressure sensitive tapes now dominate the case sealing market and huge quantities are produced. Figures quoted by C Cervellati of Exxon in Italy suggest a European production total of 2.4 billion square metres in 1988, of which 50% was made from OPP and 46% from PVC film. Solvent based natural rubber coating technology accounted for 88% of the production.

These figures, compared with previous years', confirmed two trends. One is a move to polypropylene as the preferred film substrate, the other is greater use of emulsion (water based) acrylic adhesives. The second trend is only in its infancy but predictions are that it will become much more significant as environmental pressures continue to grow.

Personalised printing of self adhesive tapes, once possible only for customers willing to order huge quantities, is now much more accessible to smaller users. The main reason for this is that acrylic based adhesives on polypropylene tapes do not pose the same release problems as rubber based adhesives on PVC. This in turn reduces the need to use toluene based inks to cut through the release coating, and allows simpler printing inks and equipment to be used, with corresponding savings in capital cost.

Printed tapes add to the security of sealed cases in many instances, and there are a number of other refinements on this security element. One is the printing of invisible messages by inks capable of being read only under ultraviolet light. Another, developed in Japan, is based on an abhesive material (to prevent adhesion) printed before it is coated with a pigmented adhesive. In use, if the tape is removed the adhesive separates selectively, leaving a readable message in the pigmented adhesive on the container or case.

MISCELLANEOUS CHEMICALS

Many chemicals are used as constituent parts of the materials mentioned elsewhere in this review. Just a few are also used on their own for specific purposes. Among the more important of these are aerosol propellants. The earliest aerosols used easily liquefied hydrocarbon gases as propellants because they had the right performance and were relatively cheap, but they were flammable. When the new family of refrigerant gases, the CFCs (chlorofluorocarbons), became available in the 1950s they offered much better properties – totally inert, non-flammable (indeed, a related group of chemicals, the halons, are used for fire extinguishers), and they were fully miscible with a wide range of liquid products. As production quantities grew they also became progressively cheaper.

By 1984 the world was producing 1.2 million tonnes per annum of these CFCs, almost all of which was allowed to diffuse into the atmosphere. Scientists had been concerned for a few years about certain thinning effects on the ozone layer high in the stratosphere which filters out much of the potentially harmful ultraviolet irradiation from the sun. By this time they were sure that there was a causal relationship between CFCs and the depletion of the ozone layer, and the pressure was on for legislation to reduce the consumption of CFCs, and for scientists to devise safer alternatives. The immediate option available was to revert to the hydrocarbons, mainly butane and propane, while the search went on, and this has happened. Under the aegis of the 1987 Montreal Protocol agreement the major world users agreed to freeze production and then progressively reduce it over the next decade. Events soon overtook this level of measured control and calls for an even faster reduction and even a complete ban on their use have since grown. There are three technical ways in which this problem can be alleviated in addition to the continuing use of hydrocarbon propellants. The first is to use related chemicals of the same family but which are not fully halogenated, ie they retain at least one hydrogen atom in their molecule. These have much reduced stability, and they break down before being able to contact the ozone layer. A second approach is to totally replace the chlorine element within the molecule by fluorine. Both of these routes are being used by various companies and new manufacturing plant to produce them is being commissioned around the world. Thirdly is the use of completely different chemicals having similar properties, dimethyl ether being the most promising of these. For food and medicinal uses nitrogen dioxide can also be used.

The nomenclature of CFCs is confusing and no single convention is universally adopted. One of the most straightforward lists, based on a publication issued by the pressure group Friends of the Earth, is as follows:

Table 5.1
Environmental impact of CFCs

Substance	Chemical formula	Ozone depletion potential
CFC 11	$CFCl_3$	1.0
CFC 12	CF_2Cl_2	1.0
CFC 113	$CCl_2F\text{-}CClF_2$	0.8
CFC 114	$CClF_2\text{-}CClF_2$	0.8
CFC 115	$CF_3\text{-}CF_2Cl$	0.4
HCFC 22	CHF_2Cl	0.05
HCFC 123	$CCl_2H\text{-}CF_3$	0.02
HCFC 124	$CHClF\text{-}CF_3$	0.02
HCFC 141b	$CH_3\text{-}CCl_2F$	0.06
HCFC 142b	$CH_3\text{-}CClF_2$	n/a
HFC 125	$CHF_2\text{-}CF_3$	0
HFC 134a	$CF_3\text{-}CFH_2$	0
HFC 143a	$CH_3\text{-}CF_3$	0
HFC 152a	$CH_3\text{-}CHF_2$	0
Methyl chloroform	$CH_3\text{-}CCl_3$	0.15
Carbon tetrachloride	CCl_4	0.2

The first five of these chemicals are the existing propellants; the next five are the so-called HCFCs with their 'available' hydrogen atoms to encourage molecular breakdown; and the rest are the fully fluorine-substituted forms. As can be seen, the ozone depletion potential decreases in the same order.

Much discussion among environmental scientists and commercial companies centres on whether it is better to move rapidly to adopt the HCFCs which are already available, with their reduced potential for ozone damage, or to push ahead on producing the HFCs, which, since they have to be extensively tested for safety before they can be widely adopted, will not be available in commercial quantity for some years.

Meanwhile another effect of this particular environmental debate has been a huge increase in the number of ideas for producing alternative pressure dispensing packs which do not use either CFCs or hydrocarbons. Among those on offer have been pump-up packs using air, compressed nitrogen dispensing containers, carbon dioxide produced in situ from acid and sodium bicarbonate, rubber compression sleeves, metal springs, and of course the whole range of mechanically operated pump dispensers and trigger sprays.

ATMOSPHERE-MODIFYING CHEMICALS

Desiccants which absorb water from the air inside a container are the most widely-used of these. They have a finite absorbing capacity (although silica gel can be regenerated by heating to dry off the absorbed water). Care must be taken when using any desiccants to ensure that the amount of moisture packed in with the product does not exceed their absorptive capacity – never forgetting that timber can hold its own weight in water and is seldom below 7-8% by weight. A second requirement is that the pack must be totally sealed otherwise the desiccant will simply draw in moist air until it is exhausted.

Other chemicals which absorb water include calcium chloride and active alumina. The first of these is seldom used in sensitive packaging situations, since it dissolves and can cause corrosion. It is, however, the standard material in the laboratory test procedure for measuring the WVTR of films or packs.

Volatile corrosion inhibitors – sometimes also called vapour phase inhibitors – are used to retard corrosion of ferrous metals in sealed packs, especially for military and engineering items. These sophisticated materials must be used with a full awareness of their mode of working, as in certain circumstances they can actually make matters worse.

In Japan there is a great vogue for what are called 'freshness agents' for food packaging. The most important are those which absorb oxygen from the headspace gas to prevent, or at least retard, oxidation reactions which lead to food spoilage. Best known is the Ageless range from Mitsubishi. This is a finely divided metallic iron which (put crudely) simply rusts and uses up the available oxygen in the process. To eliminate the unsightly rusty appearance which these can sometimes show, and also to prevent rejection problems when metal detectors are used on the packing line, some other chemicals are now being employed. Ascorbic acid (vitamin C) is one of the favoured substitutes.

Other chemicals used to inhibit spoilage include alcohol vapour which is allowed to diffuse from small sealed sachets via a thin wick, and saturated sugar solutions wrapped between two layers of very permeable film to modify the water activity of certain products such as fish, to prolong shelf life. Ethylene absorbers, active carbon, zeolites, even chemical catalysts to oxidise ethylene, and lactic acid, and others are suggested to retard the ripening of fruit, or preserve other foods. The variety of ideas and scope of applications for these (particularly in Japan) is bewildering but the underlying science is still in its infancy and only the most fully understood – mainly Ageless – have been adopted outside of Japan. This particular system was used for a ground coffee pack by General Foods of the USA in about 1986. One product of this type not originating from Japan is Smartcap, a crown cork closure developed by US company Aquanautics, for beer packaging which contains an oxygen absorber built into the liner material in the crown cork. This absorbs the oxygen from the headspace gas and is said to prolong the life and enhance the quality of bottled beer. Another, from CMB Packaging, has been described at page 110.

CUSHIONING

Compressible materials have been used for centuries to locate and protect delicate items by filling gaps and absorbing the energy if the container is subjected to any impact. Straw, wood wool, and shredded paper were the main naturally occurring materials, although the Chinese were known to use tea to protect porcelain during its export to the West in the 18th Century. The effectiveness of this last example was vividly demonstrated when the Nanking treasure was salvaged a few years ago. Recent environmental concern has led to the re-introduction of one such natural material – see page 19.

From the natural materials other more or less processed materials including rubberised hair and foam rubber were then developed. Once plastics became available, whole families of cellular materials were developed, based on polyethylene, polyurethane, polyether, EVA and polystyrene. Density and rigidity affect the energy absorbing properties of these materials and these may be calibrated to design packs appropriate to the fragility of the product, and the anticipated level of shocks. Laboratory drop hammers are used to measure the rate of deceleration provided by different materials or designs.

Expanded polystyrene is a particularly widely used cushioning material, in the form of moulded fitments, cut blocks, and loose-fill shapes. A variant on the last named offered by Dow Chemicals has the individual polystyrene granules coated with a low-tack adhesive. These then adhere, forming a solid block when compressed around the actual object being protected. Sac-choc, from Lepiney Industries in France, achieves a similar effect by evacuating a plastic bag loosely filled with cushioning granules.

Foam-in-place is another form of cushioning, produced by pumping into a container two liquids which react to produce a foam which then hardens into a semi-rigid cellular material. These are based mainly on polyurethane compounds.

Most recent have been the lightweight bubble films available in three or four different bubble sizes and heights. These provide excellent location and void-filling performance, contribute virtually no weight to the pack, are completely inert and non-absorptive, and introduce no chemicals into the pack. Their only limitation is that air is forced out of the cells if heavy weights remain on them in one position. Improved grades from Sealed Air Corporation are now available which incorporate a PVDC barrier coating to reduce this tendency. A variation on these developed in Japan is Tricone, a system in which sealed empty pouches of air are formed on site as required for cushioning and void-filling uses. A continuous tube is produced from narrow reels of heat sealable PET film. This is inflated with air and sealed at fixed intervals, each seal being rotated by 90° to produce tetrahedral configuration pouches. A particularly clever element of this system is that the air being used to fill the continuous tube is allowed to pass through a narrow orifice which cools it many degrees below the ambient level. This cold air then sealed into the pouch, expands to reach the normal ambient temperature and results in a slight but significant over-pressure inside the sealed pouches.

INKS

A theme which can be traced in various parts of this review is that each component or material used must offer as many benefits as possible; meeting more than one functional requirement can often provide the optimum economics.

Even the inks used to print the containers and labels are now available in forms which can provide at least one extra benefit in addition to conveying the printed message and graphics. Among those now available are thermochromic inks, incorporating chemicals which change colour with temperature. Some have just one colour change point, others two or more. Their widest use is in Japan, particularly the Metamo range from Toppan Printing. Some uses are merely novelty such as designs which appear on paper cups when filled with hot or chilled drinks. Others are slightly more functional such as temperature indicating labels for wine, and a level-indicating scale (activated by the chilled contents) is another used for opaque 3l beer packs.

Humidity-sensitive inks are also available. These have long been used as inexpensive test papers, but they can now be printed on labels to alert users to climatic conditions in circumstances where these may need to be closely controlled or monitored.

Related, although using more sophisticated chemicals, is the range of 'time-temperature' indicators used to monitor the storage conditions of chilled or frozen foods. In their simplest form these consist of low melting point coloured waxes which soften and spread at a fixed temperature. Others are changes in microencapsulated low melting point waxes. Both provide only a one-off point of reference, ie "This pack has at some time(s) been exposed to a temperature above.....".

For real confidence in the quality of the food distribution chains more sophisticated methods are needed. Manufacturers and retailers need to know both the temperature and the aggregate time for which a predetermined temperature level has been exceeded. A number of systems are now available to do this. Three of the systems currently available or very near to being fully commercialised provide a useful set of examples. Fortunately each also employs a different principle. 3M's Monitor Mark from the USA employs a low melting point wax and a porous wick. Each time the wax melts as it reaches a sufficiently high temperature threshold it progresses up the scale – the higher the temperature, and the longer it is maintained, the faster and farther the wax moves.

I-Point from Swedish company Biotechnology makes use of an enzyme, the rate of activity of which is proportional to temperature, in order to produce a change in the acidity of a small solution held inside a capsule on a label. This in turn changes colour due to the presence of a pH sensing indicator.

Lifeline's Freshness Monitor, also from the USA, relies upon the slow (time/temperature dependent) polymerisation of a colourless acetylinic monomer into a coloured polymer. This can even be printed as a bar code format so that at a certain level of visibility it can be read automatically.

Ultraviolet readable inks have been mentioned at page 121 above. They can be either broadband readable or, if specially formulated, respond only to precisely defined wavelengths of light in order to provide a security function.

Magnetic inks have also a reasonably long history, being originally developed for electronic reading systems. These have also some scope for use in pack security systems. One form of this, demonstrated by Dutch carton company Targa Offset, involves the attachment of a label printed with a bar code in magnetic ink on the inner surface of a carton. Its magnetic field is set by the product manufacturer and cancelled out at the store checkout. Any pack being taken past a scanning head fitted at the store exit, which had not been through the cash checkout, triggers an alarm.

Finally, a range of photochromic inks was launched at the end of 1989 by Canadian company Traqson and its UK arm, Traqmark. These incorporate microencapsulated quantities of complex chemicals which can be switched in an instant from colourless to deep red by a flash of blue light, and as quickly back again by another flash using white light. Photoswitchable inks with this speed (10^{-11} seconds) and sensitivity, have never before been available and they offer many opportunities for security coding of information on products as diverse as airline tickets, bottles of whisky or patient-specific medicines.

At the time of launching the chemicals were available in the form of litho ink and flexo inks, but screen inks and even ink jet applied types were being developed. A wider spectrum of colours was also expected to be available within a matter of months and even more sophisticated systems deriving from the research would make one-way irreversible switching a possible option. Photoswitching by very precise narrow-band wavelengths of light is another possibility. Each of these would further enhance the security opportunities.

LABEL MATERIALS

The humble label, once little more than a piece of paper stuck on to a pack, is now capable of providing many more functions than merely a carrier of information. The materials from which these are made are also selected from a much wider range.

Paper based

These remain the mainstay of the industry, but the proportions which are now self adhesive, as opposed to the traditional wet-applied plain paper label, continues to grow. In 1985, according to specialist consultancy Labels and Labelling Services Ltd, an estimated 30 billion labels were used in the UK. Of these, 60% were plain paper, 26.6% self adhesive, and 10.5% gummed paper. Ten years earlier the figures were 70%, 12%, and 16% respectively.

Other materials

Most plastics films and laminates are printable, and so may be used as label stock. Polyester, polystyrene, and oriented polypropylene are increasingly being used, in transparent, frosted, pigmented, and pearlescent forms. Their properties are constantly being modified, giving them many of the physical attributes of paper, while avoiding its hygro-instability (frequently a cause of 'curl' on application machines).

Plastics substrates used for self adhesive labels make possible invisible labels on glass or plastics containers. Polystyrene filled with mineral materials such as talc or titanium dioxide has handling properties very similar to a good quality paper. Dow Chemicals offers Opticite as one form of this. Cavitated oriented polypropylene, which also has a paper-like feel, offers benefits in cost and yield (the density is ultra low). An additional benefit is its ability to stretch to a greater degree than paper – this is important where it is used for wrap-around labels for carbonated soft drinks in PET bottles; the bottle expands slightly over time and can break a paper label which is unable to stretch to this extent.

Soluble papers offer some benefits in returnable bottle systems, while for very high performance situations, such as hazardous chemicals, tough, spun-bonded polyolefin materials, eg Tyvek, are used. Other materials for special application include edible labels made from natural collagen – used to label meat carcasses – and a shrinkable PVC/aluminium foil laminate sleeve label developed in Japan. The latter works by uniformly pleating the foil as the PVC shrinks, the laminating adhesive being applied in a series of fine lines or a grid to facilitate this.

Polyethylene sleeves are stretched mechanically and applied to large plastics drums or other cylindrical packs.

Technical developments have centred on the self adhesive in the following specific ways:

> *Use of water based adhesives, mainly the acrylic type. These allow lower cost production plant since expensive solvent extraction equipment is not required.*
>
> *Lower cost release backing materials. An example is the use of the electron beam curing technique for the silicone coating. The electron beam, since it produces less heat, allows cheaper but thermally sensitive materials such as thin HDPE, to be used as a backing material.*
>
> *Elimination of the backing material altogether. This involves coating the printed surface of the label with a release material such as silicone and leaving the labels in an uncut form on the reel. Effectively the reeled labels are similar to printed tape. On a bottling machine a die cutter is used to punch out the required shapes. Waddingtons of the UK has commercialised this as Monoweb and H J Heinz was the first major user to adopt it in 1988. The system saves money by eliminating the backing web, and by allowing longer runs between reel changes. It is most appropriate for major users where a standard shape (not necessarily standard design) of label is used.*

Another way of dispensing self adhesive labels, saving the cost of a backing, is the Solo system. Here the labels are provided pre-punched and in blocks requiring no on-machine punch. Application is by simply presenting a block of labels (adhesive side uppermost) to the container which then pulls off the top label. This too is made possible by developments in release coatings and the technology of applying and curing them.

Much of the interest in synthetic substrates for labels centres on the subject of plastics recycling. Traditional paper labelled plastics containers cannot be recycled economically since the paper contaminates the recovered plastic. Using labels of plastics which are compatible with the body material overcomes this limitation. The Diamond label substrate introduced in 1989 is one plastic offered particularly with this in mind, and the manufacturer, Revolutionary Adhesive Materials Ltd, claims that both the adhesive and the substrate are totally compatible with all commonly used plastics, allowing complete recycling.

In addition to the traditional separate adhered label two new systems are now being increasingly used, each having implications for the materials selection.

Shrink sleeve labels have been extensively used in Japan since the mid-1970s and moved to Europe in a big way about 10 years later. They comprise a clear plastic sleeve reverse printed on the inside which is slipped over the body of a container and which then shrinks (almost exclusively around the girth) to tightly grip the container when passed through a heating tunnel. Initially all were made from PVC, but today other plastics including PET, PS and OPP are also being adopted.

Benefits are that very high quality graphics are possible, the print is protected from scuffing, the bottle is also afforded some protection from surface scratches (extremely important in the case of glass), relatively complex shaped bottles can be all-round labelled, and by arranging the sleeve to grip over the closure a tamper evident feature is also achieved.

In 1989 the major producing company, Japan's Fuji Seal, also offered the sleeves in oriented polypropylene which has the advantages of lower cost and better compatibility if polyolefin bottles are to be recycled. The same principle has also been used with sleeves of expanded polystyrene foam for some years. Here the main purpose is to protect lightweight glass bottles, although other peripheral benefits claimed are quieter handling, better grip (especially when wet) and an element of thermal barrier.

Both types also offer some reduction in the velocity and hence the distribution of glass fragments in the event of a glass bottle containing a carbonated beverage being dropped. PVC sleeves taken virtually to the bottle neck are particularly effective in this respect.

A new concept, developed by Fuji Seal in conjunction with French company Ferembal in 1989, is a 'foamable coating' which is printed in narrow stripes on the inside of the sleeve label. At the time of shrinking the label to the container it is unaffected, but when exposed to higher temperatures as in a microwave or conventional oven, the coating foams up to about 200µm thickness, providing a thermal barrier to protect the fingers while handling the hot pot.

The second new form of labelling is in-mould application, in which a printed label with a heat-sensitive coating on its back surface is placed inside a mould in which a plastic container is to be produced. The hot plastic causes the label to firmly adhere over its entire area and leaves no projecting edge for possible catching. The system can be used with injection moulding, extrusion-blow moulding, injection stretch blow moulding and thermoforming. In the last named system a paper label is often used to provide an essential component of the wall stiffness. Yoghurt and similar dessert packs are very common examples of this.

Substrates may be any of the usual label materials: paper, foil or plastics, including transparent types. As with shrink sleeve systems there is concern to facilitate recycling of plastics containers. This is even more important with in-mould labels because if this is not possible the economics of the manufacturing process can be adversely affected if faulty bottles cannot be reprocessed in-plant. Furthermore, if the label and the bottle are of an identical or closely similar polymer then no heat sensitive coating may be needed and the printed image is effectively fused into the surface. Care is needed to choose a label material which is not affected by the heat of the molten or softened polymer.

Labels may also be used to carry other elements – microencapsulated aromas in printed patches, temperature or relative humidity sensing inks, and holographic images for product security.

DEGRADABLE MATERIALS

At the present time, calls for packaging materials to be made degradable, or more frequently to be biodegradable, are in vogue. The motives behind this are not entirely clear but two elements certainly loom very large in the thinking of many people. One is litter and the other is the increasing quantities of waste material generated by our current society – with frequently made cries that there are no more 'holes in the ground' in which to hide it all away.

Looked at objectively, calling for degradability of packaging materials to deal with the first of these is treating a symptom and not the cause. Litter is the fault of dirty people (mainly) plus some accidental scattering from improperly secured loads, unemptied collection bins, and scavenging animals. It looks unsightly and certainly detracts from the visual environment whether this be on the beach, an inland beauty spot or a paved town centre.

Mounting piles of domestic refuse are again a symptom of current conditions but not so easy to dismiss in scathing terms as litter. Affluence in the developed world is the underlying reason: the 'disposable' society now discards its household fittings, furniture, cars, clothes and journals more frequently, and hence generates waste in greater quantity than ever before.

Packaging of course features in all this rubbish – about 30% is the usual estimate by weight in the domestic dustbin or bag. But, with the greater use of self-delivery of garden refuse and heavy items to local authority refuse tip sites the actual proportion is probably significantly below this. Plastics materials of all types account for about 7% of the total discarded domestic refuse in both the UK and in Europe as a whole. But, due to its low density and bright appearance, also the form in which it is used such as lightweight bottles, it visually appears to be greater.

Packaging is also a major constituent of litter, as the most emotive pictures of the debris left behind after a rock concert show. This is more an indictment of the attendees and the organisers than any fault of the packaging. It is discarded by the same people who debase other forms of activity, from rooting-up trees, to graffiti, drug taking and football hooliganism: the minority exploiting modern technology and social developments to cause well-meaning people to consider restrictions and extreme cost burdens to be imposed on the majority.

That is the background, now for some of the attempts to meet the call for degradability by technological means. The first thing to recognise is that degradability is a function of time; almost everything, from granite to gold, degrades, given enough time, so what people really mean is rapid degradability. Secondly, what does degradability mean? For some materials, eg granite, it is merely a size reduction without material change, for others such as a piece of iron it is reversion to its more stable chemical state (from which man extracted it), iron oxide or rust. Organic materials, plant and animal remains, are degraded by nature in a totally different way. The building blocks or elements are rearranged by the forces of chemistry and micro-organisms "into something rich and strange" (Shakespeare).

Of the materials used in packaging, paper and board, the largest single sector, has of course a natural origin – trees – and will, given time and suitable conditions, degrade into harmless chemicals – ultimately the carbon dioxide and water from which they were originally synthesised. But it should be remembered that papyrus, which was the original form of paper, can still be found in Egypt 2,000 years or so after it was made, preserved by the dry atmosphere of a tomb. More recently, archaeological researchers in the USA, studying the rate at which materials decay in Victorian refuse tips, have unearthed paper in near perfect condition after twenty and even sixty years. A propos this last point, it is also worth mentioning that natural materials like vegetable and meat scraps were also found in good condition, demonstrating that the landfill environment is not conducive to helping nature to reabsorb even its own produce.

Metals, both aluminium and tinplate, degrade back to their oxides, the latter at the faster rate, retarded only by lacquers and painted coatings which we put on to decorate and protect them. Glass degrades very slowly – again subject to the conditions it meets. Victorian glass bottles found in rubbish tips, often have a cloudy and flaky surface, and glass can certainly be dissolved by very pure water given enough time; but as with paper, perfect specimens are found in tombs millennia old.

This brings us to the most contentious material, plastics. Being derived from a natural material (which is itself a degradation product of animal origin) one might expect these to break down spontaneously. Of course they do, but this brings us back to the question of time scale and most of the effort put into making plastics the useful materials they are today, has involved improving their durability.

When a demand seemed to arise for degradable plastics, the knowledge gained in that endeavour was then used in reverse to identify ways of making them less stable. Three main mechanisms are used to make plastics degradable: the incorporation of photo-initiator chain-breaking additives, the addition of starch to encourage micro-organisms to attack the material when it is buried, and the development of biopolymers which are inherently degradable. The first two can be used in combination.

Various forms of the first two types have been around for 10-15 years and have had some success in restricted market areas. Most have been offered for carrier bags and other non-food contact applications, and also as mulch films for agriculture. One form of photodegradable plastics based on the addition of traces of complex metallic additives was developed by Prof G Scott at Aston University in the UK. Another type functions by the inherent instability of certain links in the molecular chain. The best known of these is Eco-plastic, actually a copolymer of carbon monoxide and ethylene. This is extensively used for the production of skeleton ring carriers for multipacks of cans. In a number of US states these items must be degradable by law. HiCone is the main supplier of these carriers. The use of degradable plastics is by no means universally seen as being beneficial. Put against the main benefit which is to break down if disposed of improperly (litter) or to help conserve refuse disposal facilities, they have a number of major shortcomings:

> *They all cost more than traditional materials, partly because the additive is itself more costly, but also because segregated production is likely to be needed and recycling of in-factory waste is inhibited.*

> *With starch-filled materials in particular, some reduction in strength usually occurs and this necessitates an increase in the amount of plastics used to maintain the original performance. Thus the starch, far from being a low-cost substitute for part of the polymer content, is often an addition.*

Furthermore, the original thought that low grade surplus starch grains could be used for this purpose was not realised, and it has been found necessary in many instances to modify the starch's properties in order to make it compatible with the plastic resins. This in turn means that the starch costs much more than was originally envisaged.

Degradation performance is rarely quantified in specific terms, eg the light exposure level and time versus fall-off in physical strength, and for all kinds there has to be a compromise between reasonably rapid breakdown but without an unacceptable deterioration of physical properties during use. Most critical of all is that the two conditions – buried in soil to allow micro-organisms to attack the starch particles, and exposed to sunlight to initiate the breakup – are mutually exclusive. As one observer said "the first type is ineffective if dropped as litter and the second type if buried in a landfill".

Where attempts have been made at giving performance levels they have either been related to paper, or the results of actual tests indicate that full degradation does not take place for months or even years, hardly solving the litter problem. In the USA one piece of legislation calling for materials to be biodegradable defines this as having a breakdown rate which is "not more than 110% of the time required for paper to do so" – presumably under identical test conditions.

Breakdown which results in materials existing in smaller particles greatly increases the surface area, increasing the opportunity for oxygen and light to achieve its further breakdown, but this can also take a lot of time. Meanwhile the plastics material is still there, only in a smaller form.

The detailed chemistry of all intermediate breakdown products is not fully known – the claim for 'carbon dioxide and water only' is based on theoretical chemistry. Hence some reservations have been expressed on the desirability of these unknown intermediate materials being in the environment.

Finally, in terms of the social objective of reducing litter, there is a clear danger that by labelling materials as biodegradable this will actually be encouraged. The foregoing reservations account for the general reluctance of the plastics industry to endorse the use of degradable materials based on either form of additives, to solve social problems for which they do not see they are responsible.

A wider consensus has now (1990) developed among consumer groups and environmental activists that "biodegradable" materials are not the answer to waste reduction. They add to costs and inhibit recycling, which is a preferred option.

The third type of naturally degradable materials is quite different and is one into which much effort is now being put. The capability of forming transparent films from cellulosic chemicals has long been known – regenerated cellulose film, for instance, is degradable – although when provided with the excellent barrier protection of a PVDC layer this can take a very long time.

Other film materials have been produced from carboxymethyl cellulose and many starch derivatives. In Japan films have also been prepared from a variety of vegetable materials including orange peel, and from other natural substances such as chitin, derived from crab shells. These particular materials are also in fact edible. There are a number of research projects studying materials of this type. Fairly obviously all water soluble films, most of which are based on polyvinyl alcohol (see page 138) are also degradable, as is EVOH (see page 88).

Most interest in recent years has focussed on completely new forms of plastics, of which Biopol from ICI – a hydroxybutyrate/hydroxyvalerate copolymer – has shown the most promise. This is produced from carbohydrate solutions, mainly sugar, by a selected bacteria strain. The end product may be cast into transparent films having similar properties to oriented polypropylene, or extrusion blow moulded into bottles. Its cost is currently very high but claims from Japan, where a number of researchers are working on it, suggest that much lower costs are now within reach, if not actually achieved. As has been pointed out earlier, the demand/production rate equation usually results in a lowering of manufactured costs of such materials.

The properties can be varied by modifying the feedstock to produce different proportions of hydroxybutyrate to valerate. Results reported from the ICI company Marlborough Polymers, and the Tokyo Institute of Technology in the period 1986-89 show a rapidly growing understanding of the properties of this material and steady progress to reduce its manufacturing costs.

Small quantities of PHB have been produced and packaging in the form of films and blow moulded bottles provided to customers to evaluate. West Germany is the first country in which market trials have been carried out for toiletry products.

Also in Japan, the Shikoku Institute is working on a related but unspecified biopolymer made from a mixture of sugars. Clear films cast from aqueous solutions are claimed to be stronger than nylon. Although degradable after two to three months in soil, the film can, it is claimed, be used in the presence of water.

Warner-Lambert in the USA also developed a starch-derived plastic material Novon in 1990.

The Battelle Research Institute in Frankfurt has also produced clear films from a starch base, and a major project is in hand to develop this to a commercially viable process. Of equal and potentially much greater importance in the long term of these materials, is that they do not depend on petrochemical feedstock. They therefore provide a possible source of plastics based on completely renewable vegetable resources such as sugar and starch.

WATER SOLUBLE FILMS

Most of these are based on polyvinyl alcohol. They are widely used for the manufacture of disposable bags used in hospital laundries to reduce the possibility of cross-infection. Other special applications in packaging include the unit dose packaging of difficult or hazardous materials such as powder dyes or agrochemicals.

Edible films are a bizarre concept to some, since being part of a food or medicinal product they themselves have to be protected from contamination during handling. They do, however, offer benefits in certain areas, making possible unit doses of powders for instance. In this respect they are similar to gelatin capsules.

A dozen or so edible films have been produced over the years, mainly in the USA and Japan. Examples are: Klucel from Hercules Inc in the USA, which is based on hydroxypropylcellulose, and is available in both film and foam form (the latter may be impregnated with medicines or dietary supplements); Ediflex, derived from maize starch; Edisol, a methylcellulose-based material; Myvacet and Lepak which are acetylated monoglycerides, the former from Eastman Chemicals; and Pullulan, a starch-based film from Toppan in Japan.